MW01222020

Nondestructive Evaluation & Testing *of* Masonry Structures

Nondestructive Evaluation & Testing *of* Masonry Structures

By **Bruce A. Suprenant, Ph.D., P.E.**
and **Michael P. Schuller, P.E.**

***Reviewed and recommended by* The Masonry Society**

Nondestructive Evaluation & Testing *of* Masonry Structures

Published by The Aberdeen Group
426 S. Westgate St.
Addison, IL 60101

Editor: Desiree J. Hanford
Art Director: Jill Taylor Whitson
Technical Editor: Richard Henry Atkinson
Copy Editor: Chris Santilli
Cover Design: Charles A. Kacin
Publisher: Mark DiCicco

 Library of Congress Number: 94-43389

10 9 8 7 6 5 4 3 2 1

ISBN 0-924659-57-2

Item No. 4530

Library of Congress Cataloging-in-Publication Data

Suprenant, Bruce A.
Nondestructive evaluation and testing of masonry structures / by Bruce A. Suprenant and Michael P. Schuller.
p. cm.
"Reviewed and recommended by the Masonry Society."
Includes index.
ISBN 0-924659-57-2 : $89.95
1. Masonry—Testing. 2. Non-destructive testing. 3. Buildings-—Testing. I. Schuller, Michael P., 1964- . II. Masonry Society (U.S.) III. Title.
TA435.S836 1995
693'.1—dc20

94-43389
CIP

ABOUT THE AUTHORS

Bruce A. Suprenant is group vice president and editorial director at The Aberdeen Group, a publishing company for the construction industry. He has a B.S. in construction from Bradley University, an M.S. in structures from the University of Illinois in Urbana, and a Ph.D. in civil engineering from Montana State University. His practical experience includes working as technical director for CTC-Geotek Inc., a materials engineering and testing firm in Denver, Colo. He was formerly an associate professor at the University of Colorado, Boulder.

Dr. Suprenant is a member of the Masonry Standards Joint Committee, The Masonry Society (TMS), the American Society for Testing and Materials (ASTM), the American Concrete Institute (ACI), the American Arbitration Association, and the American Society of Civil Engineers (ASCE). He is past chair of the Executive Committee of ASCE's Materials Engineering Division.

Dr. Suprenant is a registered professional engineer in five states. His industry experience includes working as a structural engineer for Sverdrup & Parcel in St. Louis, Mo., and as an analytical structural engineer with the Portland Cement Association in Skokie, Ill.

Michael P. Schuller is a consulting engineer with Atkinson-Noland & Associates Inc. in Boulder, Colo., specializing in evaluation and repair of unreinforced masonry buildings. He has a B.S. and M.S. in civil engineering from the University of Colorado, Boulder. He has worked on many commercial projects regarding building evaluations with clients that include the Nuclear Regulatory Commission, U.S. Army Corps of Engineers, University of Georgia, National Park Service, and private building owners.

Mr. Schuller has published more than 20 papers relating to evaluation, repair, and structural behavior of masonry structures and served on the organizational committee of two recent international conferences on nondestructive evaluation of civil structures and materials. Mr. Schuller is a registered professional engineer and a member of the American Society of Civil Engineers (ASCE), The Masonry Society (TMS), and the Association for Preservation Technology (APT), and is involved with the development of test standards for nondestructive evaluation of masonry with the American Society for Testing and Materials (ASTM) and RILEM. He received the Alan H. Yorkdale Memorial Award in 1991, presented by The American Society for Testing and Materials (ASTM), for the best paper pertaining to masonry during that year.

CONTENTS

Chapter 3 • *Evaluating Cracks*

Chapter 4 • *Strength Testing*

Chapter 5 • *Surface Techniques*

TABLES

FIGURES

Introduction

A large percentage of today's buildings are constructed of masonry. These buildings range in age from a few years old to hundreds of years old. The safety of these buildings under extreme loads from earthquakes and wind, as well as the safety of existing buildings when they are used for a different function, requires detailed condition assessment studies. In the past, this was often accomplished by removal of prisms for laboratory testing — a procedure that both disfigures the building and is subject to considerable error.

The application of nondestructive evaluation (NDE) methods for masonry have been developed in the last 15 years. These methods are able to provide significantly more reliable and useful information than that supplied by destructive prism tests, which impact the building's appearance. In the case of historic or important masonry buildings, these methods are especially attractive.

The authors have written a comprehensive treatise on the use of nondestructive evaluation methods for masonry buildings. The first few chapters cover the very important first step of planning the overall evaluation and the initial requirement for an organized visual condition survey, including crack mapping. Following chapters describe available nondestructive methods in detail, including when a technique should be used, the equipment required, how to conduct the test, and how to interpret the data. A valuable appendix is provided, listing equipment manufacturers for the various NDE methods discussed.

R.H. Atkinson, P.E., Ph.D.
Boulder, Colorado
October 1994

Chapter One • *Planning the Evaluation*

1.1 *INTRODUCTION*

Much time, effort, and expense can be wasted on in-place testing unless the objectives of the investigation are clearly defined. The objectives of the investigation affect the choice of test methods, the number and location of the tests, and the interpretation of the results. Careful and thorough planning minimizes the investigation's cost, effort, and time schedule and ensures the final results are useful.

In-place testing of existing masonry structures rarely proceed as planned. The initial investigation program, based on project objectives, should be continually reviewed as test results are obtained. When appropriate, update the investigation program based on the available data.

Understanding the range of tests available and their limitations and accuracies enables the engineer to plan an investigation that provides useful test results. Besides understanding the tests, engineers should recognize the influence of the operator on the final test results. Although some tests appear relatively simple, trained technicians working under the direction of an experienced supervisor should perform the tests and experienced engineers should interpret the results.

Poor planning can result in disappointing and uninformative results and frustration with the nondestructive equipment. Poor planning also requires the engineer to be more creative and inventive in analyzing the results and forming conclusions. Avoid disputes regarding tests and test results with a well-planned investigation.

1.2 *OBJECTIVES OF IN-PLACE TESTING*

The objectives of in-place masonry testing can be grouped into three main categories: quality control, compliance, and evaluation. Quality control testing ensures an acceptable supplied and constructed product. Prisms may be removed from a structure to confirm the acceptance of the materials and workmanship. Compliance testing verifies proper construction. For instance, radiography has been used to verify reinforcing lap splice lengths in grouted masonry walls when the contractor's personnel were observed to be hand stabbing rebars into fresh grout. Evaluation testing attempts to predict the performance of the structure or a structural element. Nondestructive testing can be used as input to a structural evaluation that assesses the integrity or safety of a structure. The evaluation may be required because of a change of occupancy, purchase, material deterioration, or structural damage.

In-place masonry testing of repairs also is important. The acceptance and verification of repairs to masonry structures, especially for seismic retrofit, have become more prevalent in recent years.

Engineers usually determine the objectives of in-place masonry testing. Owners, architects, building occupants, and others provide constraints that must be considered when developing the investigation program to meet the desired objectives. Limited access, budget restrictions, architectural esthetics, building code requirements, liability and legal concerns, and time stipulations must be blended into a well-planned investigation program.

1.3 *AVAILABLE TEST METHODS*

There are many test methods for the in-place evaluation of masonry. A brief summary of some principal test methods used to evaluate masonry is provided in Table 1. These tests can be used to check the specific masonry problems or conditions as shown in Table 2.

Not all test methods are nondestructive; some require extraction or removal of specimens. Some leave indentations, grease spots, holes, or other marks that might be architecturally unacceptable. If the test or repair where a sample was extracted might not be acceptable to the architect, use the test method in an inconspicuous location. The architect can then observe the effect of the test on the building esthetics.

Extraction of test specimens such as cores usually requires water to cool the core barrel while drilling. This water may damage exterior and interior surface finishes. Also, other tests may generate dust, noise, radiation, and debris that must be considered while testing. Utilities, such as electricity, gas, and compressed air, and access items, such as scaffolding, forklifts, and hoists, are inherent to some test methods.

Some test methods provide a direct measurement of strength. Other test methods must be correlated with prisms or cores to provide a strength estimate. Don't force a test method to provide information on strength. Tests, like rebound and penetration tests, are more suitable to indicate the uniformity of the in-place masonry.

1.4 *EXAMINATION AND SAMPLING OF MASONRY STRUCTURES*

1.4.1 *Examination*

There is an ASTM C 823 Standard for the Examination and Sampling of Hardened Concrete in Constructions [6] but not for masonry. Many engineers rush into getting a test result then wonder what the test result means. An understanding of ASTM C 823 as it could pertain to masonry can benefit the planning of an investigation program.

ASTM C 823 divides the examination into a procedural plan, preliminary investigation, assembly of records, and finally a detailed investigation. The *procedural plan* establishes that the examination of in-place masonry is to be undertaken according to the scope, objective, and systematic procedures. When appropriate, the procedural plan should be agreed to by all responsible parties. The plan should stipulate the expected budget and time for the planned examination.

TABLE 1. TESTS USED TO EVALUATE MASONRY

Visual-optical: Includes visual inspection for cracks, weathering, mortar deterioration, corrosion, efflorescence, and other similar defects that can be detected without the use of optical aids such as low-power magnifiers. Also includes measurement of differential structural movements and use of fiber optics to detect internal cracks, voids, and flaws.

Rebound: Use of a spring-driven steel hammer or a pendulum hammer to determine the uniformity of in-place masonry and to delineate zones or areas of poor-quality or deteriorated masonry. Can be used on individual masonry units, mortar joints, or masonry assemblages.

Pullout: Measures the force required to pull out an embedded anchor and the surrounding masonry. The anchor is epoxied into a masonry unit or mortar joint or sometimes a helical anchor is drilled into a mortar joint. The measured pullout force is usually related to other strength tests but can be used to assess material uniformity. Pullout tests can also be used as proof tests of anchors embedded into existing masonry.

Drilling: Measures the energy required to drill a bit into a mortar joint. Used to determine the uniformity of mortar joints and to delineate areas of deteriorated mortar.

Penetration: Determines the resistance of masonry units or mortar to penetration by steel probes or pins. Used to determine the uniformity of masonry units and mortar joints.

Push (Shove): Measures the sliding shear strength of a masonry bed joint by displacing a single masonry unit horizontally with a hydraulic jack (a flatjack can also be used in the adjacent head joint). The adjacent masonry unit and head joint must be removed to provide room to insert the hydraulic jack and to allow movement of the unit to be tested. Flat jacks are sometimes used to control or adjust the normal pressure applied to the masonry unit that is being pushed. A standard test method is described in the *Uniform Code for Building Conservation* [1].

Core: Measures the shear-bond between a masonry unit and a bed joint by loading an 8-inch diameter in compression (load applied 15 degrees from the diametral bed joint). Another core test measures the shear-bond between a masonry unit and grout by using a guillotine apparatus to apply a direct shear force along the bond line. A standard test method is described in the Los Angeles Building Code [2].

Bond Wrench: Measures the flexural strength by applying a bending moment to a torque wrench that is clamped onto a masonry unit adjacent to the designated test mortar bed joint. Masonry units above the tested unit must be

TABLE 1, *continued*

removed to attach the bond wrench. A standard method for laboratory testing is ASTM C 1072-86 [3].

Flatjack: Measures the compressive stress in the masonry by inserting a flatjack into a slot cut into a mortar bed joint and increasing the pressure until the original distance between points above and below the slot is restored. By inserting two flatjacks into parallel slots, then pressurizing, the stress-strain properties of the masonry are measured. A standard test method for both compressive stress and masonry deformability are provided by ASTM C 1196-91 [4] and C 1197-91 [5].

Pulse Velocity: Measures the time of travel of a pulse or train of waves through masonry to determine the uniformity, to indicate changes in masonry properties, or to survey structures to estimate the severity and extent of deterioration, cracking, and voids.

Impact Echo: Measures the depth of a reflecting surface for a stress wave generated by mechanical impact. Detects and delineates internal discontinuities in masonry and, with interpretation, identifies the nature and orientation of the discontinuities.

Magnetic: Uses a portable magnetic test device to locate rebars, ties, and anchors. Also measures the distance from the masonry surface to the rebar and the size of the rebar.

Radiographics: Uses X-rays or gamma rays primarily to determine the location of reinforcing (includes ties and anchors). Can also be used to detect voids or other flaws.

Infrared Thermography: Uses selective infrared frequencies to identify heat patterns characteristic of certain defects. Has been used to locate bands of mortar droppings, voids in solid masonry walls, and location of wall ties.

Corrosion Activity: Detects active corrosion by measuring the half-cell potential.

Petrography: Uses microscopic examination, sometimes in combination with other techniques, to evaluate samples of mortar, grout, and masonry units. Features that can be evaluated include air content, carbonation, bond, mortar ingredients and quantities, coatings, contaminants, and adequacy of brick firing.

Load Testing: Applies a test load to a structure or structural element that simulates the design loading. Serviceability criteria such as water leakage, excessive deflection, cracking, or even structural failure are monitored and measured visually or with detection devices. Typical load tests include lateral wind loading, proof loading of new or existing anchors, lateral loading of piers, and free vibration.

TABLE 2. MASONRY TEST METHODS AND APPLICATIONS

	Test Method for Investigating Condition																
Condition	Visual-optical	Rebound	Pullout	Drilling	Penetration	Push (Shove)	Core	Bond Wrench	Flatjack	Pulse Velocity	Impact Echo	Magnetic	Radiographics	Infrared Thermography	Corrosion Activity	Petrography	Load Testing
In-place strength?		X	X		X	X	X	X	X	X	X						X
In-place uniformity?	X	X	X	X	X				X	X							
In-place deformability?									X								X
In-place stress?									X								
Crack location?	X									X	X		X				
Crack movement?	X																
Performance under load?																	X
Rebar size, location, cover?												X	X	X			
Anchor and tie location?												X	X	X			
Voids in grout?	X									X	X		X	X			
Voids in masonry?	X									X	X		X	X			
Corrosion of rebar?	X														X		
Durability problems?	X									X	X					X	

The purpose for undertaking masonry investigations is substantially the same as that suggested by ASTM for concrete investigations:

- To determine the ability of the masonry to perform satisfactorily under expected conditions of future service
- To identify the processes or materials causing distress or failure
- To discover conditions in the masonry that caused or contributed to unsatisfactory performance or to failure
- To establish methods for repair or replacement without hazard or recurrence of the distress
- To determine conformance with construction specification requirements
- To develop data to aid in fixing financial and legal responsibility for cases involving failure or unsatisfactory service
- To evaluate the performance of the components used in the masonry structure

The planned investigation program should produce an explanation concerning the masonry condition that was the purpose of the investigation.

The *scope of the investigation* may be limited to isolated areas showing deterioration, cracking, water intrusion, or excessive deflections. Or the scope of study may include a large portion or even the entire structure. The scope may not only include masonry but also include the foundations, diaphragms, and other structural elements that comprise the complete masonry building.

The *preliminary investigation* consists of the condition survey. This part usually includes a visual inspection, review of project plans, specifications, testing and inspection reports, and occasionally the use of simple test methods. Most engineers prefer to break a project investigation into preliminary and detailed phases. The preliminary investigation provides the prerequisite information to develop a well-planned detailed investigation program. Also, it's easier to provide a cost estimate for the detailed investigation once the preliminary investigation is complete.

After the preliminary investigation is complete, the engineer undertakes the *assembly of records*. The assembly or review of these records may include the assessment of environmental loads including wind, rain, snow, temperature, ice, or earthquake. Maintenance or operation records might indicate the possible cause or beginning of a problem. Interviews with the contractors, occupants, or neighbors also might help in developing an information database.

A *detailed investigation* of masonry should include all procedures that are required to achieve the approved scope and objectives within the authorized budget and time schedule. The detailed investigation usually includes (a) a thorough examination of the masonry, (b) surveys (photographs, drawings, crack maps) and field tests to define and evaluate the condition of the masonry, and (c) samples for field or laboratory testing.

A *report* is produced that details the investigation, observations, procedures, and conclusions. If the report is for a legal inquiry, then the recommendations of ASTM E 620 [7], E 678 [8], and E 860 [9] might apply. These standards recommend practices for evaluating data and reporting opinions for investigations that are undertaken for potential litigation.

1.4.2 *Sampling*

For in-place evaluation of masonry structures, material samples can be extracted from the structure for testing. Samples are of two types: (a) those intended to represent the variability of the masonry structure and (b) those that display specific features of interest (such as cracking, spalling, weathering) but are not intended, individually or collectively, to represent any substantial portion of the masonry structure.

It's important that samples at areas of specific interest, usually distressed areas, are considered separately from samples that represent the entire structure. Sometimes, especially in litigation, investigators focus on obtaining samples and test results in areas that feature cracking, spalling, or general deterioration. Unfortunately, the investigators occasionally use these results to make assessments about the quality of the masonry or workmanship for the entire structure. Be careful; a good attorney can get your results thrown out of court if you've inadvertently combined samples from areas of specific distress. This doesn't mean you can't include test results from cracked or spalled areas. You can include these test results but only if a random sampling plan chooses the sample at that cracked or spalled location.

A random sampling plan is required when choosing samples or test locations that are to represent the entire structure. The sampling locations should be spread randomly or systematically over the entire structure. A masonry structure is usually divided into lots of 5,000 square feet of wall area each. A lot is chosen by random and then a test location within that lot must be randomly selected. ASTM E 105 Standard Recommended Practice for Probability Sampling of Materials [10] provides guidance on developing a probability sampling plan.

A sampling plan not only chooses random locations but it also must indicate the sample size. ASTM E 122, Standard Recommended Practice for Choice of Sample Size to Estimate the Average Quality of a Lot or Process [11], provides guidance for choosing a sample size. The Uniform Code for Building Conservation [1] provides some guidance on selection of test locations for in-place shear tests:

> ***Location of tests.*** *The shear tests shall be taken at locations representative of the mortar conditions throughout the entire building, taking into account variations in workmanship at different building height levels, variations in weathering of the exterior surfaces, and variations in the condition of the interior surfaces due to deterioration caused by leaks and condensation of water or by the deleterious effects of other substances contained within the building. The exact test location shall be determined at the building site by the engineer in charge of the structural design work. An accurate record of all such tests and their location in the building shall be recorded and these results shall be submitted to the building department for approval as part of the structural analysis.*

The sampling plan also must cover procedures: who takes the samples and how are they extracted, labeled, protected, transported, and tested. There are few standardized tests for existing masonry. The testing laboratories employed for the tests should have a written procedure to train their employees. The engineer should review and approve the test procedure. The test procedure must describe any limits on the rate of loading, eccentricity of loading, and uniformity of loading. Don't assume that all testing laboratories perform each test listed in Table 1 in the same manner.

1.5 *CHOOSING SAMPLE SIZE*

The number of samples to test for new construction is usually specified by the applicable building code. Some codes provide criteria for the number of samples to test for existing construction (see Table 3). However, the choice of the number of samples to test for an existing building is quite often left to the engineer. ASTM E 122, Standard Recommendation Practice for Choice of Sample Size to Estimate the Average Quality of a Lot or Process [11], provides some guidance in the selection of sample size.

TABLE 3. SUMMARY OF IN-PLACE TEST SAMPLING REQUIREMENTS FOR MEASURING SHEAR STRENGTH OF UNREINFORCED MASONRY

Sample Requirement	*Test*	*Required Test Value*	*Reference*
1. One test for each combination of different grout type and/or masonry unit combination per 5,000 sf of wall area, but not less than one per building.	Core test for shear bond	Average unit shear strength, maximum load divided by shear area, shall not be less than 100 psi.	1984 ATC [16]
2. At each of both the first and top stories, not less than two per wall line or line of wall elements. At each of the other stories, not less than one per wall element providing a common line of resistance. In any case, not less than one per 1,500 sf of wall area and a total of eight.	In-place shear tests (Push tests or testing 8-inch-diameter cores for shear)	The minimum quality of mortar in 80 percent of the shear tests (push test) shall not be less than the total of 30 psi plus the axial stress in the wall at the point of test. The mortar joint tested in shear (core test) shall have an average ultimate stress of 20 psi based on the gross area.	1991 Los Angeles Building Code [2]
3. Same as in No. 2.	In-place shear test (push test)	Mortar shear test values, *v*, in psi shall be obtained for each in-place shear test: $v_{to} = (V_{test} - P_{D+L})/\text{Area}$. Individual unreinforced masonry walls with *v* consistently less than 30 psi shall be entirely pointed before retesting. The mortar shear strength, v_t, is the value in psi that is exceeded by 80% of all the mortar shear test values, v_{to}. Unreinforced masonry with mortar shear strength, v_t, less than 30 psi shall be removed or pointed and retested.	1991 Uniform Code for Building Conservation [1] and 1992 San Francisco Building Code [17]

The equation for the size, *n*, of the sample is:

$$n=\left(3 * \frac{\sigma}{E}\right)^2 \tag{1-1}$$

where:

σ = the advance estimate of the standard deviation of the lot

E = the maximum allowable error between the estimate to be made from the sample and the result of measuring all the units in the lot

3 = a factor corresponding to a low probability that the difference between the sample estimate and the result of measuring all the units in the lot is greater than *E*. The choice of 3 is recommended for general use and makes it almost certain that the sampling error will not exceed *E*. Where a lesser degree of certainty is desired, a smaller factor may be used.

Factor	Approximate Probability
3.00	3 in 1000 (0.003)
2.58	10 in 1000 (0.010)
2.00	45 in 1000 (0.045)
1.96	5 in 100 (0.050)
1.64	10 in 100 (0.100)

It's often more convenient to use the above equation as:

$$n=\left(3 * \frac{V_0}{e}\right)^2 \quad (1\text{-}2)$$

where:

V_o = the advance estimate of the coefficient of variation expressed as a percent

e = the allowable sampling error expressed as a percent

These equations can be used to estimate the number of in-place tests to perform on an existing building. However, as ASTM E 122 mentions, the allowable cost for the project may dictate testing a smaller sample size.

1.5.1 *Calibration*

What is a reasonable allowable sampling error for masonry? By calibration from existing codes, a sampling error can be computed. For instance, the 1991 Uniform Building Code [12] and the 1992 Specifications for Masonry Structures (ACI 530.1) [13] requires three prism tests for each 5,000 square feet of wall area. The Canadian Masonry Design Standard [14] and the RILEM Test Standard [15] requires a minimum of five prism tests. Thus, the number of sample sizes of prism tests for new construction is set by codes.

The coefficient of variation for making and testing prisms in the laboratory is about 6%. The coefficient of variation for making prisms in the field and then returning them to the laboratory for testing is about 10% (based

TABLE 4. SAMPLING ERROR OF PRISM TESTS

	Coefficient of Variation (%)			
Sample Size	6	10	15	20
3	10.4	17.3	26.0	34.6
5	8.0	13.4	20.1	26.8
8	6.4	10.6	15.9	21.2

on test data from CTC-Geotek Inc., Denver [18]). Table 4 shows the sampling error for prism tests given the sample size and the coefficient of variation.

A sampling error of ±17.3%, three prism tests with a coefficient of variation of 10%, is the approximate value provided by following the U.S. codes for sample size and a good-quality testing program.

For predicting the number of other in-place tests, a sampling error of 20% is a reasonable value to assume. Another procedure, however, is to relate the number of in-place tests to provide the same degree of testing certainty (that is, about a 17% sampling error) as three field prism tests.

1.5.2 *Number of In-place Tests*

From equation 1-2, assuming the same maximum allowable error and the same probability factor, the ratio of the number of in-place tests to the number of the standard prism tests equals the ratio of the squares of the corresponding within-test coefficient of variations. This equation can be used to provide the same degree of testing certainty for an in-place test as three prism tests.

$$\frac{n_{in\text{-}place}}{n_{prism}} = \left(\frac{V_{in\text{-}place}}{V_{prism}}\right)^2 \qquad (1\text{-}3)$$

where

$n_{in\text{-}place}$ = number of in-place tests
n_{prism} = number of prism tests
$V_{in\text{-}place}$ = within-test coefficient of variation of in-place tests
V_{prism} = within-test coefficient of variation of prism tests

TABLE 5. ESTIMATED COEFFICIENT OF VARIATIONS, %

Technique	*Within-test COV Lab Testing*	*Batch-to-Batch (Overall) COV Field (Testing & Materials)*
Prism Test	6/10%*	20%
Flat Jack Test	10%	18%
Push (Shove) Test	23%	30%
Core Shear Test	28%	40%
Bond Wrench	30%	36%
Ultrasonic Pulse Velocity		
Direct	8%	20%
Indirect	12%	30%

* 6% for prisms constructed and tested in the lab and 10% for prisms constructed in the field and tested in the lab

Typical values for the coefficient of variation for different in-place masonry tests are shown in Table 5. The lab testing values are provided for within-test COV's and the field data represents typical batch-to-batch (overall) COV's. The within-test COV for the prism test is shown as 6% for prisms constructed and tested in the lab, and 10% for prisms constructed in the field, transported, cured, and tested at the laboratory.

By using the within-test coefficient of variation reported in Table 5 and equation 1, the number of tests equivalent to three prism tests per 5,000 square feet can be determined. Table 6 lists the number of in-place tests for each test method that provides the same degree of certainty as testing three prisms for every 5,000 square feet. However, many other factors such as cost, access, importance of the structure, and perceived quality of the in-place masonry must be considered by the engineer when choosing sample size.

TABLE 6. NUMBER OF TESTS TO EQUAL THREE PRISM TESTS PER 5,000 SQUARE FEET

Test	Number of tests
Flat Jack Test	3 tests
Push (Shove) Test	16 tests
Core Shear Test	24 tests
Bond Wrench	27 tests
Ultrasonic Pulse Velocity	
Direct	3 tests
Indirect	4 tests

Note that building codes require the same number of in-place shear tests for both the push and the core test. The coefficient of variation of each test indicates the push test is more reliable than the core test.

1.6 *IN-PLACE MASONRY VARIABILITY*

In-place masonry properties will vary within a structure, within a member, and even between adjacent masonry units and mortar. The construction of masonry structures is an art, the properties of which are highly dependent on the skill of the workers. Masonry assemblages are constructed of individual units bonded together by batches of job mixed mortar. Expect a minimum overall coefficient of variation of 10% for any masonry property even for the best constructed masonry.

Figures 1-1 through 1-4 illustrate the in-place variability of masonry properties. The average, high, and low ultrasonic readings for a grouted brick wall are shown in Figure 1-1. A variation of 2,000 feet per second, the difference between high and low readings, for direct transmission ultrasonic readings are common along the wall [19]. The variation of in-place flexural bond strength is shown in Figure 1-2 [20]. Figure 1-3 shows the variation of direct transmission ultrasonic pulse velocity plotted as velocity contours for walls with different applied vertical stress [21]. Figure 1-4 indicates the variation in masonry deformability within a masonry structure measured in-place using flatjacks [22].

Figure 1-1

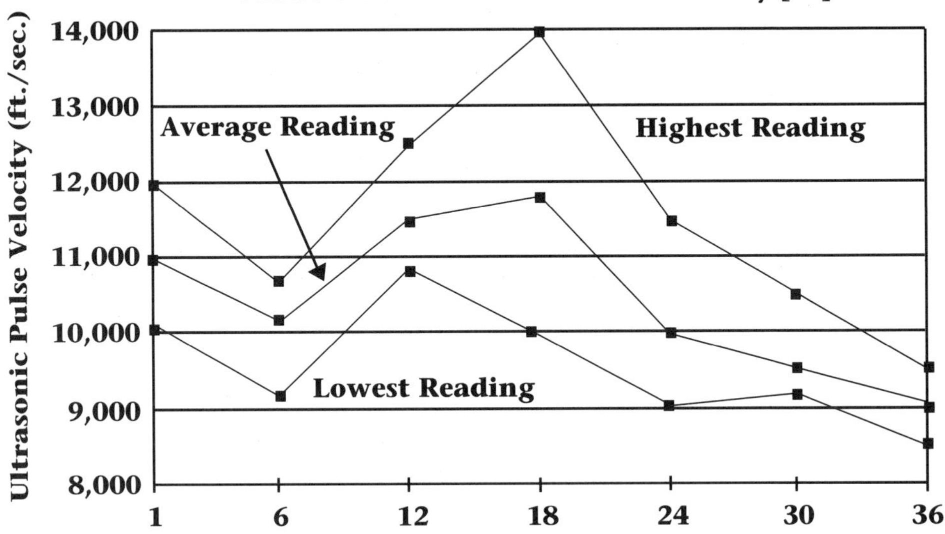

Figure 1-2

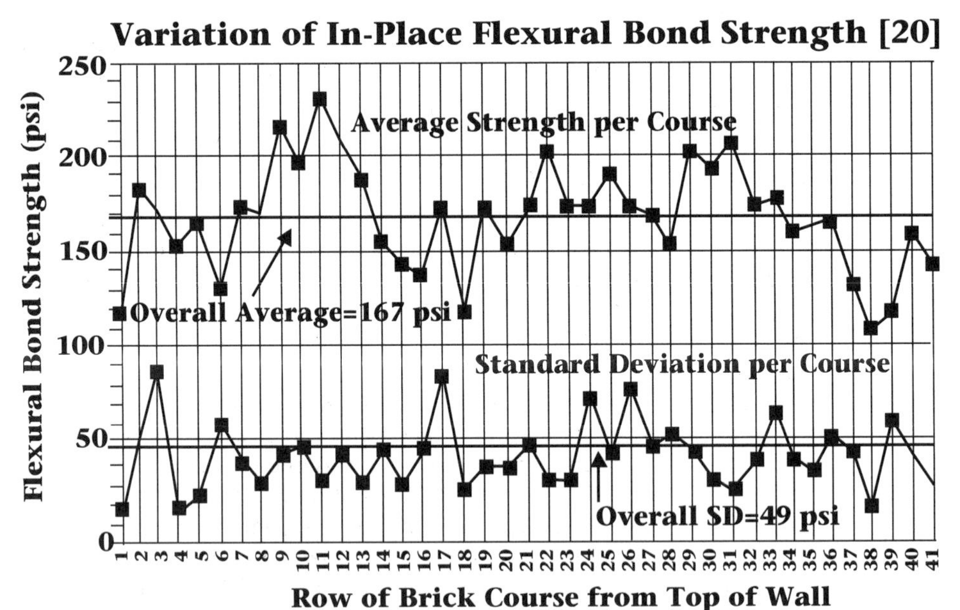

Figure 1-3. Variation of ultrasonic pulse velocity (ft/sec) measured for different levels of vertical stress on wall E1.

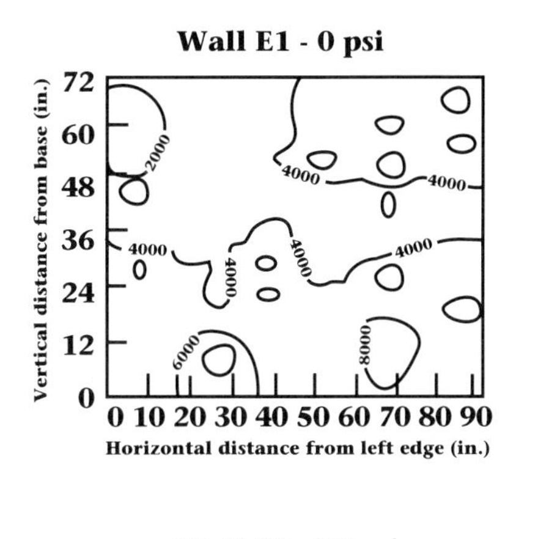

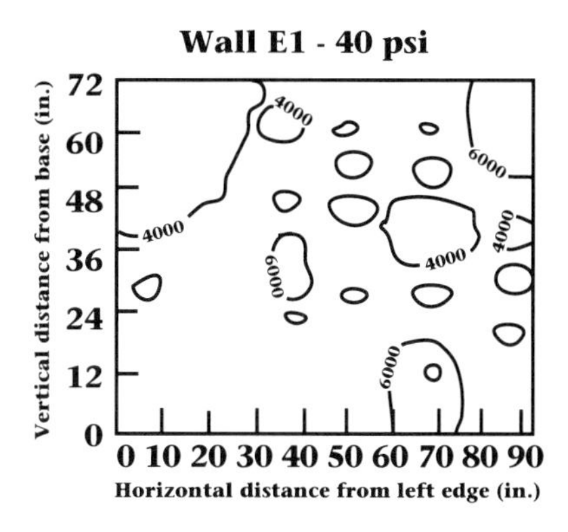

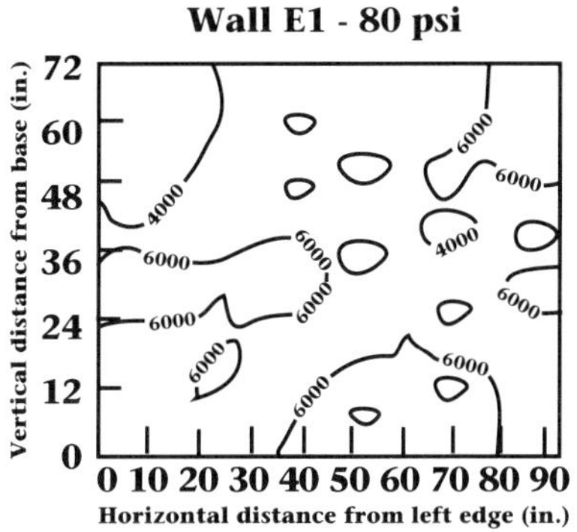

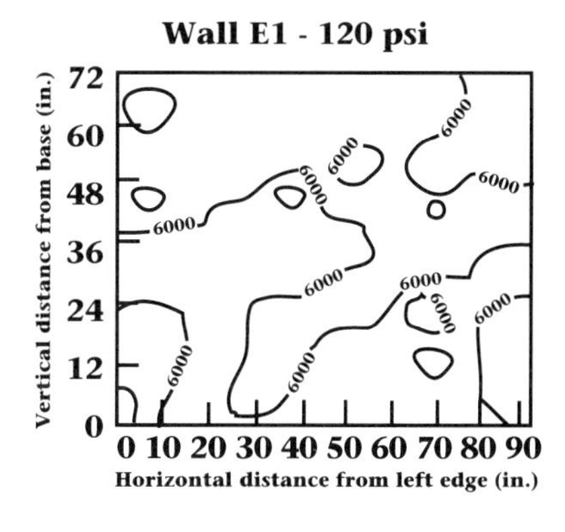

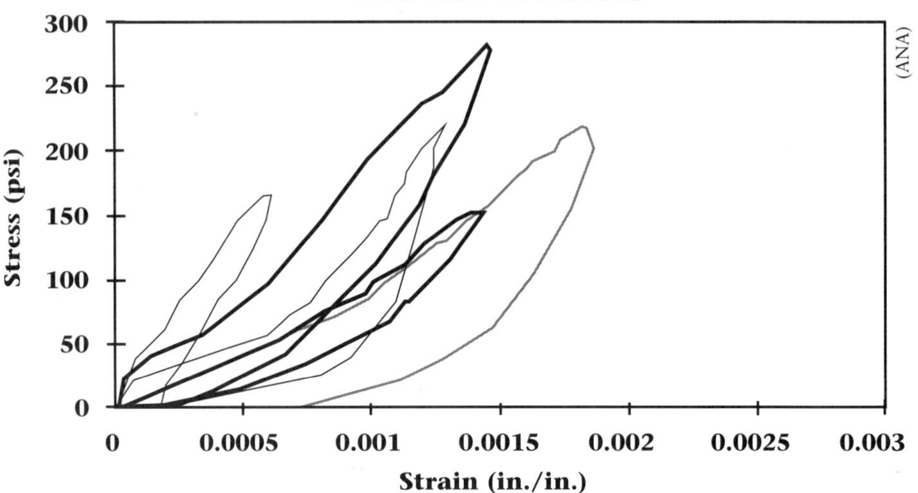

Figure 1-4. Five separate in-place deformability tests, all conducted on the same structure, illustrate the variability in material properties that can be expected.

These figures show how masonry properties vary. Because the workmanship is a major factor in masonry property variation, it's difficult to predict any predetermined pattern in masonry properties. On occasion, however, the middle of each day's work appears to be better than the top or bottom of that days work. Masonry is typically stronger when workers lay units at waist level. The workers spread the mortar better and there's less wiggling and tapping on the units to make them level.

Workers laying units while bending over or when stretching don't always produce the same quality. Sometimes, looking at the masonry wall reveals where the scaffolding was raised. You can tell when the workers had to be reaching up to try to lay the last course. A tower scaffold that moves as work progresses keeps the workers laying units at waist level, thus minimizing the effects of bending or reaching.

Although workmanship is usually the main factor affecting masonry variability for new construction, there are other factors that influence the masonry variability of existing structures. Differences in the applied vertical stress produce different measured in-place shear stress values (push test). Masonry assemblages near the bottom of a wall, where the dead load is the greatest may test stronger. The applied vertical dead load can minimize wind-induced microcracking while masonry in parapets don't benefit much from applied vertical compression.

Environmental factors also play a role in the variability of masonry within a structure. In most locations there is a preferential rain and wind direction that adds moisture and loads the structure in a prescribed pattern. Wind and rain can be more of a problem along edges, cornices, or parapets. Freeze-thaw cycles are more prevalent on a south facing wall than on the north wall. Even without visible damage, test results on a south facing wall may be as much as 30% lower than test results from a wall facing north.

1.7 IN-PLACE STRENGTH RELATIVE TO STANDARD SPECIMENS

Prisms and cores are sometimes extracted from a structure to measure the ultimate compressive stress or the ultimate shear stress. A height-to-thickness correction factor must be applied to a prism. But what about a correction for curing, age of specimen, or effect of specimen removal? For masonry, there isn't much of a need for correcting for differences in curing or age of specimens. The suspected loss of strength from damage caused by specimen removal has always been a guess.

A recent study [23] conducted by the research and development laboratory of the National Concrete Masonry Association shows the relationship between a prism sawcut from an existing wall and a prism constructed of the same materials. The results of this study are presented in Table 7. Note that in two cases the prisms sawcut from the walls were stronger than the constructed prisms. The sawcut ungrouted brick masonry was more than 1000 psi stronger than the corresponding constructed prisms. One consistent feature was that the sawcut prisms all had a coefficient of variation that was higher than the constructed prisms.

Although the study recommends that three sawcut prisms be equal to three constructed prisms, this really isn't the case. To obtain about the same degree of testing certainty (see equation 1) as three constructed prisms, a minimum of five sawcut prisms should be tested. For example, grouted concrete masonry would take 15 tests of saw-cut prisms to provide the same degree of testing certainty as three tests of constructed prisms.

TABLE 7. SUMMARY OF PRISM TEST RESULTS

Prism Type	Average Net Area Compressive Strength (psi)	Standard Deviation (psi)	Coefficient Of Variation (%)	Sawcut divided by Constructed
Ungrouted Concrete				
Sawcut	2215	183	8	0.94
Constructed	2366	153	6	
Grouted Concrete				
Sawcut	2843	315	11	1.06
Constructed	2678	312	5	
Ungrouted Brick				
Sawcut	5803	688	12	1.23
Constructed	4699	402	9	
Grouted Brick				
Sawcut	4085	740	18	0.95
Constructed	4280	436	10	

1.8 INTERPRETATION OF RESULTS

Use standard statistical tests to interpret in-place test data. It's not sufficient for the average in-place strength to be greater than the calculated applied load. The interpretation of results must account for the uncertainties that exist. Some test methods measure the strength of masonry directly and others, such as ultrasonic pulse velocity, must be correlated with a strength value. A common requirement (see Table 3) is to base the in-place strength on the 20th-percentile value. Because of small sample size, there is a confidence level associated with using the 20th percentile strength values.

The engineer must decide whether the nondestructive evaluation test should be performed before or after the masonry sample has been extracted. Performing the test after extraction is easier, but the test may be performed on an area damaged by sample extraction and transportation. The engineer should consider testing the sample before and after extraction to compare the values and test procedures.

1.8.1 Direct Strength Measurement

The American Concrete Institute Committee 228 report "In-Place Methods for Determination of Strength of Concrete" [24] recommends some approaches for the calculation of in-place strength. One simple approach is outlined below. Because of the high variability associated with in-place masonry testing, a more rigorous calculation approach is not yet justified.

In this approach the average and standard deviation of the sample tests are calculated. The 20th-percentile masonry strength is obtained by subtracting the standard deviation times a factor (which varies with the number of tests and desired level of confidence) from the sample average. The factors used in this approach are one-sided tolerance factors for data that are normally distributed [25]. The values for the factors for different numbers of tests and confidence levels are provided in Table 8.

Suggested values [24] of the confidence level are 75% for ordinary structures and 90% for important structures such as medical or other facilities that must remain open during an earthquake. A confidence level of 95% has been suggested for nuclear power plants. ACI 228 indicates that a confidence level of 75% is widely used in practice. Unfortunately, only the 80% values are listed in literature for the 20th percentile.

The example in Table 9 illustrates the procedure. Nine cores were removed from a masonry building and loaded in compression and the shear stress was determined. What is the 20th-percentile of the ultimate shear stress?

1.8.2 Strength Correlation

Some prisms are usually extracted from masonry buildings. Using rebound hammer, ultrasonic pulse velocity, or other techniques, readings can be taken on the prisms. The prisms are then tested and the compressive stress calculated. By regression analysis the nondestructive readings are correlated to the prism tests. Further nondestructive readings on the building can then be used to estimate the masonry's compressive strength at the testing locations.

Stone, Carino, and Reeve [26] have indicated the limitations of this approach: (1) the correlation relationship is presumed to have no error, and (2) the variability of the masonry strength is assumed to equal the variability of the in-place test results. The first factor tends to make the estimates of the in-place 20th-percentile strength unconservative, and the second factor tends to make the estimates too conservative. Stone and Reeve [27] present a more rigorous approach that incorporates a regression analysis that includes the variability of both the X and Y variables.

TABLE 8. ONE-SIDED TOLERANCE FACTORS FOR 20TH PERCENTILE LEVEL

Number of Tests	Confidence Level		
	80%	90%	95%
2	3.417	5.049	6.464
3	2.016	2.671	3.604
4	1.675	2.373	2.983
5	1.514	2.145	2.683
6	1.417	2.012	2.517
7	1.352	1.923	2.407
8	1.304	1.859	2.328
9	1.266	1.809	2.265
10	1.237	1.770	2.220
11	1.212	1.738	2.182
12	1.192	1.711	2.149
13	1.174	1.689	2.122
14	1.159	1.669	2.098
15	1.145	1.652	2.078
16	1.133	1.637	2.059
17	1.123	1.623	2.043
18	1.113	1.611	2.029
19	1.104	1.600	2.016
20	1.095	1.590	2.004
21	1.089	1.581	1.993
22	1.082	1.572	1.983
23	1.076	1.564	1.973
24	1.070	1.557	1.965
25	1.065	1.550	1.957
30	1.043	1.523	1.924
35	1.026	1.502	1.900
40	1.023	1.486	1.880
50	0.993	1.461	1.852
∞	0.842	1.282	1.645

TABLE 9. EXAMPLE OF 20TH-PERCENTILE SHEAR STRENGTH

Individual Test Results	Calculation
98 psi	Mean = 122 psi
128 psi	
155 psi	Standard Deviation = 25 psi
156 psi	
94 psi	Factor (CL = 0.80%) = 1.266
129 psi	
100 psi	20th percentile strength
138 psi	
100 psi	=122-1.266*25=90 psi

The approach presented here is essentially that reported by Bickley [28]. This approach, while subject to the limitations described by Stone, Carino, and Reeve, should be satisfactory for the analysis of masonry test results.

Example. Five prisms were removed from a masonry building. The prisms were then tested using indirect ultrasonic pulse velocity, after which the prisms were tested. A regression analysis was performed between the ultrasonic pulse velocity and the prism strength. Later, eight more ultrasonic readings were taken on a part of the structure. What is the estimated compressive strength of the masonry at the location of the new ultrasonic testing? Table 10 provides a summary of the calculations.

Figure 5 shows the graphical relationship between correlating the nondestructive test with the compressive strength of prisms removed

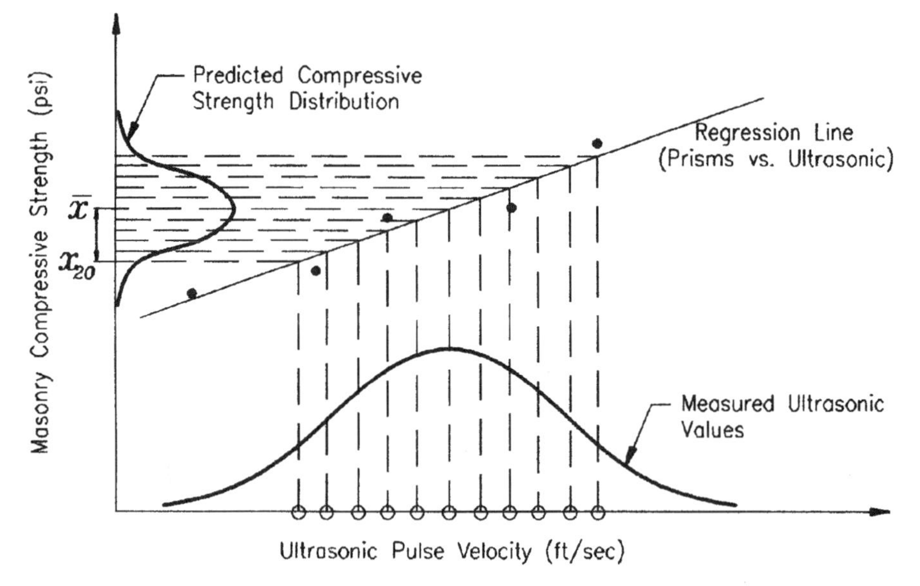

Figure 1-5. Graphical representation of computing 20th percentile compressive strength.

TABLE 10. EXAMPLE OF STRENGTH CORRELATION CALCULATION

Indirect Ultrasonic Pulse Velocity (ft./sec.)	Prism Strength (psi)
7,027	2130
8,190	3473
8,770	4046
8,900	4546
9,419	6316

Regression Equation: Mean Prism Strength = -9333 + 1.588*Ultrasonic
Lower 95% Prism Strength = -17,779 + 0.594* Ultrasonic

	Indirect Ultrasonic Pulse Velocity (ft./sec.)	Predicted Mean Prism (psi)
	6,125	393
	6,600	1147
	7,100	1941
	7,325	2298
	7,545	2647
	7,650	2814
	7,975	3330
	8,125	3568
Avg	7,306 ft./sec.	2267 psi
SD	679 ft./sec.	1078 psi
CV	9.3%	47.6%

One-sided Tolerance Limit for Prism Compressive Strength

from the wall. When tests at other locations on the wall are performed, the regression equation previously developed is used. Note that each test must be transferred to a compressive strength. Then the mean and variance of the compressive strength can be calculated and the 20th percentile value can be found.

Don't take the mean of the nondestructive tests and use the regression equation to find the mean compressive strength. This provides no estimate of the variability of the compressive strength.

1.9 *COMBINED TEST METHODS*

The use of more than one test method can increase the confidence level of the results, improve calibration accuracy, and reduce the cost of the investigation. Some test methods such as rebound, penetration, or sounding (listening to the sound of a hammer blow hitting the masonry) are used as a preliminary assessment of the homogeneity of the masonry. Also, test methods that locate reinforcing or embedded electrical conduit provide useful information before sawing into a wall.

The statistics of combining test methods can't be presented here. There are some good references indicating the use and analysis of combined nondestructive test methods for concrete [29,30].

1.10 REFERENCES

1. International Conference of Building Officials. 1991. Seismic Strengthening Provisions for Unreinforced Masonry Bearing Wall Buildings. Appendix, Chap. 1 in *Uniform Code for Building Conservation*. Whittier, Calif.: International Conference of Building Officials.
2. City of Los Angeles. 1992. *City of Los Angeles Building Code, Section 8809*. Los Angeles.
3. ASTM. 1986. *Standard Method for Measurement of Masonry Flexural Bond Strength*, ASTM C 1072-86. Philadelphia: ATSM.
4. ASTM. 1991. *Standard Test Method for In Situ Compressive Stress within Solid Unit Masonry Estimated Using Flatjack Measurements*, ASTM C 1196-91. Philadelphia: ASTM.
5. ASTM. 1991. *Standard Test Method for In Situ Measurement of Masonry Deformability Properties Using the Flatjack Method*, ASTM C 1197-91. Philadelphia: ASTM.
6. ASTM. 1988. *Standard Practice for Examination and Sampling of Hardened Concrete in Constructions*, ASTM C 823-88. Philadelphia: ASTM.
7. ASTM. 1985. *Standard Practice for Reporting Opinions of Technical Experts*, ASTM E 620-85. Philadelphia: ASTM.
8. ASTM. 1984. *Standard Practice for Evaluation of Technical Data*, ASTM E 678-84. Philadelphia: ASTM.
9. ASTM. 1982. *Standard Practice for Examining and Testing Items That Are or May Become Involved in Products Liability Litigation*, ASTM E 860-82. Philadelphia: ASTM.
10. ASTM. 1975. *Standard Recommended Practice for Probability Sampling of Materials*, ASTM E 105-75. Philadelphia: ASTM.
11. ASTM. 1979. *Standard Recommended Practice for Choice of Sample Size to Estimate the Average Quality of a Lot or Process*, ASTM E 122-79. Philadelphia: ASTM.
12. International Conference of Building Officials. 1991. *1991 Uniform Building Code*. Whittier, Calif.: International Conference of Building Officials.
13. American Concrete Institute. 1992. *Specifications for Masonry Structures*, ACI 530.1-92/ASCE 6-92/TMS 902-92. Detroit: ACI.
14. Masonry Design for Buildings. 1984. CAN3-S304-M84. Canada.
15. RILEM. 1990. *Determination of the Compressive Strength of Small Walls and Prisms*, RILEM LUM B.1.
16. Applied Technology Council. 1984. *Tentative Provisions for the Development of Seismic Regulations for Buildings*.
17. City of San Francisco. 1992. Seismic Strengthening Provisions for Unreinforced masonry Bearing Wall Buildings. In *San Francisco Municipal Code (Building Code)*, Part II, Chap. 15. San Francisco.
18. CTC-Geotek Inc. 1984. Private communication. Denver.
19. Hobbs, B., and S.J. Wright. 1986. Ultrasonic Testing for Fault Detection in Brickwork and Blockwork. Presented at Proceedings on International

Conference on Structural Faults and Repair.
20. Shrive, N.G., and D. Tilleman. 1992. As Simple Apparatus and Method for Measuring On-Site Flexural Bond Strength. Presented at Proceedings of 6th Canadian Masonry Symposium. Canada.
21. Abrams, D.P., and G.S. Epperson. 1989. Evaluation of Shear Strength of Unreinforced Brick Walls Based on Nondestructive Measurements. Presented at Proceedings of 5th Canadian Masonry Symposium. Canada.
22. Schuller, M.P. 1992. Masonry Evaluation at the Seney-Stovall Chapel. Atkinson-Noland & Associates Inc. Boulder, Colo.
23. National Concrete Masonry Association, Research and Development Laboratory. Research Evaluation of the Compressive Strength of In Situ Masonry. Herndon, Va.: National Concrete Masonry Association.
24. American Concrete Institute Committee 228. 1992. In-Place Methods for Determination of Strength of Concrete. Part 2 in *ACI Manual of Concrete Practice*. Detroit: ACI.
25. Hahn, G.J., and W.Q. Meeker. A Guide for Practitioners. In *Statistical Intervals*. New York: Wiley-Interscience.
26. Stone, W.C., N.J. Carino, and C.P. Reeve. 1986. Statistical Methods for In-Place Strength Prediction by the Pullout Test. *Proceedings ACI Journal*. Sept.-Oct., vol., 138. no. 5.
27. Stone, W.C. and C.P. Reeve. 1986. New Statistical Method for Prediction of Concrete Strength from In-Place Tests. *Cement, Concrete, and Aggregates*. Summer, vol. 8, no. 1. ASTM.
28. Bickely, J.A. 1982. The Variability of Pullout Tests and In-Place Concrete Strength. *Concrete International*. April, vol. 4, no. 4.
29. Malhotra, V.M., and N. Carino. 1991. *Nondestructive Testing of Concrete*. CRC Press.
30. Malhotra, V.M. 1984 *In Situ/Nondestructive Testing of Concrete*, SP-82. Detroit: ACI.

Chapter Two • *Conducting A Condition Survey*

A crack here, a water stain over there, weak mortar joints: what does it all mean? These clues can indicate poor maintenance, heavy loads, poor construction, or just severe environmental weathering. Finding and interpreting the clues and assembling them into a rational hypothesis is the basis for the evaluation of existing structures. Investigators have many tools at their disposal, but they must understand their use and limitations. Also, to match the structure's importance, the techniques and tests must be tailored to obtain information on the structure's strength and durability within the owner's budget. The condition survey, which includes the initial site visit and field and laboratory testing, guides the investigator to the cause of the structures problem and the appropriate remedy to that problem.

2.1 *THE CONDITION SURVEY*

To correctly identify the structure's strengths and weaknesses, if any, certain information about the structure is required. Usually this information is obtained by site visits, talking with the owner and original project team (architect, engineer, and contractor), reviewing the drawings and specifications, and utilizing both field and laboratory testing. Condition survey checklists are often useful in helping guide an investigator through an evaluation.

Not every item in the checklist should be done or can be done on every structure. The investigator must decide on the purpose of the investigation which then dictates both the testing and budget. An owner who is attempting to buy a masonry structure is looking for a different evaluation than an owner who wants to know why a structure has water leaks. And after an earthquake, a building owner wants to know if the structure is safe for occupancy or what repairs are needed to make it safe. In addition, when evaluating a historic structure the owner may not allow any material samples to be removed from the building therefore it requires special nondestructive testing.

2.1.1 *Field and laboratory testing*

Both field and laboratory testing are usually part of any condition survey. The field survey may be as simple as checking and monitoring crack locations and size, using a knife to probe for weak mortar, spraying the wall with a garden hose to see how the water soaks in or repels and dries out, or using a hammer to hunt for hollow sounds in the masonry walls. However, in conjunction with these simple tests, other more sophisticated tests using ultrasonic pulse velocity, flat jacks, or drilling and coring equipment may be required.

The laboratory testing may be of prisms or cores removed from the wall to be tested in compression, shear, flexure, or tension. Other physical and chemical laboratory testing also may be performed on mortars, grouts, and coatings from small specimens removed from the structure. Sometimes the field and laboratory testing is complex and only a few specialized consulting firms or testing laboratories may be able to perform these tests.

2.1.2 *Don't forget a background check*

The investigator should obtain the original drawings and specifications, test laboratory and inspection reports, and the contractors daily field log. Not all these items, however, may be available. Conversations with the contractor, architect, inspector, contractor, or workers may provide useful information about the materials and the construction practices used for the job. Learning how the mortar was mixed (by hand with mortar mixer, preblended in on-site silos, or ready mixed mortar delivered to the site), how the grouting was done (high-lift or low-lift, cleanouts, pumping, consolidation), and the weather conditions (hot or cold weather construction techniques implemented) can help the investigator form a rational explanation for the cause of cracking or deterioration.

2.2 *SAMPLING IS THE KEY TO INTERPRETATION*

The accurate determination of in-place strength and durability requires careful sampling. If the sample is not representative of the portion of wall under investigation, then the conclusions derived from the analysis of the samples will be wrong. Two sampling conditions are found: (1) the masonry in the structure is believed to be of similar composition and quality, and (2) the masonry in different areas of the structure may be of different composition or quality.

For similar composition and quality throughout, random sampling should be spread over the entire structure and the results analyzed to determine if they may be grouped together. For dissimilar areas or portions, random sampling should be conducted within each portion of the structure. These test results should not be combined unless it is shown by statistical tests that there are no significant differences between the means and standard deviations of the results from each area.

2.3 *MAKE THE MOST OF WHAT YOU HAVE*

Inexperienced investigators usually have trouble evaluating existing structures because they always want more data. Unfortunately, most owners do not have an unlimited budget and the scope of work must fit within that budget. Therefore, the investigator must carefully choose both the tests to be used and the samples to be tested. For a limited investigation, it also is wise to keep the owner informed about the variability of the tests and the accuracy of your answers. Don't try to impress the owner with the accuracy of your conclusions; you are paid for a realistic assessment. Part of that assessment should be the reliability of your conclusions. And who knows, once the owner understands the relationship between accuracy and budget, you might find yourself with a supplemental budget.

2.4 *GUIDE TO DETERMINING MASONRY PROBLEM CONDITIONS*

The items presented below aren't always necessary for most projects. This guide to conducting a condition survey for masonry can be useful for inexperienced engineers in the masonry field or for design professionals evaluating in-place problems. Engineers should modify and adapt any guidelines for a condition survey to fit their project, problems, scope of work, and budget.

Note: This Guide for Determining Masonry Problem Conditions (Appendixes A through E) was originally proposed by subcommittee of a national standards organization. The document was rejected because committee members felt that such a document would be used by attorneys as a legal requirement for every masonry investigation. As the introduction to this chapter states, every investigation provides differences in budget, scope, and objectives. This chapter provides only guidelines and is not intended to serve as a legal mandate to engineers. The authors wish to thank the head of the subcommittee for providing permission to use the document.

A.0 *SCOPE*

A.1 This practice recommends a method for assembling and recording information for assessing problems of walls of structures constructed with brick, structural clay tile, or concrete masonry units.

A.2 This standard is not applicable to paving, retaining walls, or stone masonry.

A.3 "Problem," as used in this Standard, is defined as observed conditions of unwanted or unexpected behavior of masonry such as leaking, efflorescence, cracking, spalling, staining, bulging, warping, or deterioration. It is generally used throughout this standard as singular but may be applicable to one or more problems.

B.0 *APPLICABLE DOCUMENTS*

ASTM Standards: See Appendix E

ANSI Standards:

- A10.8 Scaffolding, Safety Requirements for
- A10.9 Concrete Construction and Masonry Work, Safety Requirements for

C.0 *SUMMARY OF PRACTICE*

C.1 The practice includes several checklists of pertinent information that should be acquired in a comprehensive investigation of masonry problems.

C.2 The user may select those lists or items that are applicable to the specific project under investigation and should modify the scope of the lists according to the scale of the project.

D.0 *SIGNIFICANCE*

D.1 This standard provides an orderly method of obtaining and organizing information that may be required by the investigator to provide a basis for determining the cause of masonry problems or recommending corrective action.

E.0 *IDENTIFICATION OF STRUCTURE*

E.1 Name and address of structure.

F.0 *DESCRIPTION OF PROBLEM WORK*

F.1 Identification of problem as reported.
- F.1.1 Background of signs of problem.
- F.1.2 Significant dates such as beginning and completion of construction, masonry unit manufacture, and first observation of problem.

F.2 Classification of problem mode.
- F.2.1 Collapse.
- F.2.2 Cracks.
- F.2.3 Displacement or column change.
- F.2.4 Unit masonry surface deterioration.
- F.2.5 Mortar deterioration.
- F.2.6 Staining or discoloration.
- F.2.7 Water penetration or permeance.
- F.2.8 Air leakage.

F.3 Ancillary building elements exhibiting problem.
- F.3.1 Sealants.
- F.3.2 Corrosion, distress, or displacement of built-in connecting metal.
- F.3.3 Condition of lintels and shelf angles.
- F.3.4 Distortion of frames in masonry openings.
- F.3.5 Foundation movement or cracking.

G.0 *INTERESTED PARTIES*

G.1 List of names, addresses, phone numbers, and relation to project.

G.2 Include as many as possible such as owner, architect, engineers, attorneys, building maintenance manager, contractor, subcontractors, foremen, materials suppliers, and others who have, or had, an interest in the construction problem.

H.0 *PERTINENT DOCUMENTS REVIEWED*

H.1 Contracts
- H.1.1 Architect/owner agreement, owner/contractor agreement, pertinent subcontractors' Agreements.

H.2 Contract Documents.
- H.2.1 General, supplementary and special conditions.
- H.2.2 Drawings.
- H.2.3 Specifications.
- H.2.4 Addenda.

H.3 Construction Documents.
- H.3.1 Change orders.
- H.3.2 Shop drawings.
- H.3.3 Samples.
- H.3.4 Product literature.
- H.3.5 Progress reports.
- H.3.6 Minutes of progress meeting.
- H.3.7 Construction log.
- H.3.8 Construction photographs.
- H.3.9 Inspectors' reports.

H.4 Design Data
- H.4.1 Assumptions.
- H.4.2 Calculations.

H.5 Litigation Documents.
- H.5.1 Allegations and claims.
- H.5.2 Interrogatories.
- H.5.3 Depositions.

H.6 Public Documents.
- H.6.1 Applicable or governing codes such as building codes, labor laws, and ordinances.
- H.6.2 Weather records.
- H.6.3 Environmental data.

H.7 Technical Data.
- H.7.1 Test data.
- H.7.2 Literature search.
- H.7.3 Experts' reports.

H.8 Miscellaneous
- H.8.1 Photographs.
- H.8.2 Correspondence.
- H.8.3 Media accounts.

I.0 *DESCRIPTION OF STRUCTURE*

The following information should be obtained from documents itemized in paragraph 8.2.

I.1 Type and use, i.e., Commercial; Shopping Center.

I.2 Plan shape, i.e., Rectangular.

I.3 Site Location.
- I.3.1 Orientation.

I.4 Plan Dimensions.

I.5 Height and Number of Stories.

I.6 Type of Structure, i.e., Load Bearing, Steel or Concrete Frame.

I.7 Construction Details of Wall Exhibiting Distress (i.e., cavity wall with concrete masonry backup between floors and face brick on shelf angles at alternate floors).

I.7.1 Reference drawing numbers, revision dates, and other pertinent information contained on title block of the drawings used to obtain the above information.

J.0 *EXPLORATION: GENERAL*

J.1 Exploration consists of visual observations and physical exploration at the site.

J.2 A preliminary visit should be made to ascertain the logistics for future exploration and the equipment required. A cursory tour around and through the building will enable the investigator to plan the physical aspects of the investigation, estimate the time required, and establish a schedule. Check such items as availability for visual inspections, location of means or access, places for storage of equipment and samples, provisions for staging or rigging and location of adjoining or adjacent buildings.

J.3 Document each site visit as to date, time of arrival and departure, weather conditions, and personnel present and their relation to the building.

K.0 *VISUAL OBSERVATIONS*

K.1 Visual observations consist of examining and measuring the problem areas and recording the information obtained on drawings, by photography, and in writing.

K.2 The minimum equipment for visual observations include measuring instruments, camera, binoculars, magnifying glass, pad and writing instruments, or tape recorder. See Appendix A for a more comprehensive list.

K.3 Drawings.

K.3.1 Use scaled drawings or sketches to record information obtained and to store and document the information gathered.

K.3.2 Prepare or obtain drawings at a convenient scale (not less than 1/8 inch = 1 foot, 0 inch) of each elevation of the building where the problem is evident. Include all surfaces such as both sides of parapet walls. Identify the location, scale, date, and elevation. Record overall dimensions at top and bottom and those of significant elements such as openings, floor-to-floor heights, and courses. Indicate column lines, floors, and ceilings. Indicate materials. Provide sections at changes of plane.

K.3.3 Document observations of distress by use of symbols. Identify symbols. Indicate location, extent, and description. Hatching or shading is recommended; color coding does not reproduce on duplicating equipment.

K.4 Photographs.

K.4.1 Photographs should be taken using color film to document the area of distress, the visual characteristics such as extent, size, deviations from true planes, stains, and variations in color and texture, as well as the omission of elements such as weep holes and flashing. Photographs also should indicate the environmental conditions existing at the time of the survey.

K.4.2 The log or photography record should include the name and address of building, date, and orientation of each photograph. It also may include time photograph was taken, camera make and model, film used, and name of photographer.

K.4.3 When areas are accessible, an identifying card containing the building name, orientation, location, identification of building element, and other pertinent information may be included in the photo. Appropriate markers also may be used to write this information on the element being photographed. Where required for clarification, a scale, rule, or other item of fixed dimension should be included in every close-up photograph. See Appendix B for a comprehensive log and identification system.

K.4.4 When areas are inaccessible, photographs should be taken by either a zoom lens or a succession of wide angle and telescopic lenses so that each close-up can be identified as to location by the previous photograph.

K.4.5 Cameras equipped with data backs also are useful providing the background will provide sufficient contrast to enable the information from the data back to be read.

K.4.6 In addition to photographs of surface problems, include salient topographic and environmental conditions, long shots of the entire building for identification, and the step-by-step process of physical exploration as specified in section L. Photographs taken from the roof at 30-degree horizontal intervals may be useful in assessing surrounding conditions.

K.5 Measurements.

K.5.1 Measurements should be made of building elements and particularly of those that exhibit problems. Measure length of each wall at top and bottom. Plumb walls to determine deviations from vertical and horizontal alignment. Measure width and depth of joints, coursing, and weep hole spacing. Measure crack widths along the crack length and measure their depth where possible. Measure lateral displacement of elements on each side of crack. Appropriate instruments for measuring include rules, scales, plumb lines, levels, feeler gauges, transits, lasers, and calipers. Measurements of deviations from true lines and those of problem areas or elements should be recorded on the drawings or contained in the written report.

K.6 Written Report.

K.6.1 Base statements on facts, avoiding hypotheses or conclusions unless specifically requested by client.

K.6.2 Record the information from visual observations on a report form. See Appendix C. Required information should be supplemented with notes regarding tactile impressions such as damp, loose, and soft.

L.0 *PHYSICAL EXPLORATIONS*

L.1 Physical explorations consist of probing and removing portions of the building for observations and, if necessary, sampling and testing to confirm or refute a hypothesis generated by information obtained from visual observations.

L.2 Probing.

L.2.1 Probe sealant joints to verify if sealant is applied over a void. Probe weep holes to determine their depth. Probe cracks and weep holes over flashing for upstanding leg. Typical probes include pointed tools such as awls or ice picks and flat blades such as knives, feelers, and hacksaw blades.

L.3 Removals.

L.3.1 Removals should be made on a piece-by-piece basis, with great care to avoid destroying underlying conditions that should be observed.

L.3.2 As underlying materials are exposed, they should be measured and their location, with respect to other materials and building elements, also should be measured and recorded on a drawing.

L.3.3 Photographs should be taken of the undisturbed areas first, followed by successive photographs as removals progress.

L.3.4 In addition to the photographs, a written record of observations should be made. Include such items as material and condition of flashing, deterioration of mortar, lintels out of level, and corrosion of ties.

L.4 Samples.

L.4.1 Sampling consists of removing portions of the buildings that may range from a small piece of mortar to a large section of a wall.

L.4.2 Elements designated for removal must be carefully selected to yield optimal information. The number and size of samples should provide a reasonable representation of the element or structure under investigation.

L.4.3 For proper sampling, the investigator should be well equipped. Appendix A contains a list of equipment normally required for investigations. If sections of walls are to be removed, scaffolds, boatswain's chairs, or boom buckets may be required, as well as temporary utilities to operate them. Protection and insurance also should be considered. Provisions should be made for either patching and repair or the use of temporary enclosures.

L.4.4 Items to be tested should be removed in sufficient sizes and quantities to satisfy the test requirements for samples.

L.4.5 Samples should be documented. Each sample should be clearly labeled and entered into a log. Labels should state building name, date and time of sample removal, location on building, and description of element(s) removed. This should be supplemented with a photographic record.

L.4.6 Information obtained from the exposure of concealed elements should be recorded in writing and by sketches to supplement the photographs.

L.4.7 Samples should be carefully packed for shipping or storage. Sealed containers are required when moisture content is to be determined or products of corrosion identified.

L.4.8 Samples should be removed and transported with due regard to the safety of the building occupants and surrounding visitors.

L.5 Testing.

L.5.1 Testing may be performed in-place or in the laboratory, or both, including water permeance and absorption tests, chemical and petrographic analysis, determination of moisture content, strain release, and strength.

L.5.2 Tests should be selected as they pertain to the signs of a specific problem. Materials should also be checked to verify their conformance to the specifications and recognized standards. Note that some physical properties of masonry materials may undergo changes when in use. See ASTM C 67 Section 1 for further information.

M.0 *INTERVIEWS*

M.1 Interviews should be conducted with parties listed in paragraph G.2. Information from personnel who were engaged in the original construction should be solicited.

M.2 Appendix D contains typical questions that may be asked during interviews.

2.5 APPENDIX A

EQUIPMENT FOR SITE INVESTIGATION OF MASONRY PROBLEMS

A. *The following is generally provided by the investigator:*

1. Camera: (35 mm with minimum 210 mm telephoto, wide angle, and micro lenses or equivalent zoom, color film, slides, or infrared; flash attachment).
2. Photographic record forms such as index cards.
3. Magnifier with 0.1 mm measuring scale.
4. Compass to show directions and position of structure.
5. Small level.
6. Two-way radios.
7. Flashlight.
8. Tape recorder.
9. Notebook and clipboard with graph paper.
10. Waterproof felt-tipped pens, marking crayon, or chalk. Two colors each.
11. One 12-inch scale with 0.1-inch divisions. One 6-inch minimum rule.
12. Hard hat, steel-toed shoes with rubber soles, rain suit, gloves, wind breaker.
13. Sample collection bags sufficient to hold a masonry unit with specimen identification tags.
14. Calculator.
15. Heavy-duty pocket knife.
16. First-aid kit.
17. Extra batteries for electrically powered equipment.
18. Plumb bob with 100 feet of cord.
19. Calipers.
20. Binoculars.
21. Metal detector.
22. Fiber scope.
23. Masking tape.

B. *The following is generally provided by a local contractor:*

1. Gasoline-driven portable masonry saw with a minimum of three 12-inch-diameter blades.
2. Respirator dust mask.
3. Mason's chisels, 1 3/4 inches wide, three required.
4. 2# sledgehammer.
5. Braided nylon mason's line, 170 lb. test, 50 feet minimum.
6. Extruded aluminum pole with a cross-sectional area of 2 square inches by 12 feet long.
7. Aluminum I beam mason's level, 40 inches long or a telescoping dry wall layout level.
8. Scaffolding: boatswain's chair, tubular steel sections, cherry picker with bucket, swing scaffolds as appropriate for the project. Scaffolding should conform to ANSI A10.8 and sections 3 and 12 of ANSI A10.9.

2.6 APPENDIX B

PHOTOGRAPHS

1. Use color film, preferably 35 mm slides. Alternatively, Polaroid® cameras may be used. Take views as close as practicable to illustrate the object in adequate light.
2. Print information on an item identification card with water-resistant markers on a contrasting background. If the card cannot be included conveniently in the photo or the printed information will not be legible, use a log and record the information on the slide margin or back of each print.
3. Prints should be enlarged sufficiently to make the identification card legible without magnification.
4. Recommended format for the identification card follows:
   a. Building name.
   b. Address, city, state.
   c. Object identification (type of distress).
   d. Wall orientation/story/bay/quadrants (for interior faces of exterior walls and partitions use story/room identification/quadrant.) Identify quadrants clockwise as a, b, c, and d.
   e. Time/month/day/year/temperature/precipitation.
   f. Photographer.
   g. Compass direction of camera/declination from vertical.
   h. Film/ASA #/F stop/speed.
   Example:
      a. Forest School
      b. 7 Main St., Orange, NJ
      c. Cracked Brick
      d. E/2/ a-7 to a-8/ b
      e. 09:15 7/14/82 75F/Rain
      f. J. Jones
      g. Southeast - Up
      h. EKT DI/64 f4.5/125

2.7 APPENDIX C

INSPECTION REPORTING FORM

Building Identification ______________________________Inspection Date

______________________________Inspector

Defects
Inspection Item

____Noted

____Description & Location

____Remarks

ITEMS UNDER CONSIDERATION

Masonry Units Deteriorated

Mortar Deteriorated

Extensive Bond Loss Between Units & Mortar

Masonry Misaligned at Control Joints or Corners

Masonry Displaced or Misplaced

Masonry Bulging Between Control Joints

Sealant Condition at Control and Other Joints

Masonry Cracked

Foundation Corners Spalled

Signs of Water Penetration

Signs of Efflorescence

Remarks:

Parties Present:

2.8 APPENDIX D

OUTLINE FOR INTERVIEWING CRAFTSMEN EMPLOYED DURING ORIGINAL CONSTRUCTION

Seek interviews with masonry foremen, working masons, and mortar makers; ask permission to tape interview. Make written notes of pertinent answers. If answers are vague, rephrase the question.

Record the following data:

1. Name and address of the structure on which masonry problem is being investigated.
2. Interview date, time, and place.
3. Interrogator's name(s) and address(es).
4. Respondent's name, address, age, and occupation.
5. Start and finish dates of on-site employment during construction.
6. How many masons were usually employed on the job?
7. Names of any other masons working on the site. Where are they now?
8. Any labor problems at site with any trade? Specifically masonry? Was any mason fired? Why?
9. Weather protection provided for masons and masonry during construction: heaters, blankets, enclosure, wind breaks, heating of masonry units/sand/water.
10. Use of mortar admixtures, type and brand name.
11. Wall covered every night?
12. Any masonry repointed or torn down and rebuilt during original construction period?
13. Did masons use western stringing method or pick and dip method of laying brick?
14. How were head joints filled? Shoved, buttered, or slushed?
15. Was collar joint filled? Slushed, grouted, or parged?
16. If a cavity wall was built, was the cavity kept clean? How? Bed joints beveled or board on a string?
17. Was joint reinforcement used or individual wall ties?
18. Was joint reinforcement cut at expansion and control joints?
19. Were wall ties set perpendicular to plane of wall or diagonally?
20. Were bed joints furrowed or beveled? Heavily furrowed?
21. What were materials and volumetric proportions of the mortar mix?
22. Were mortar materials measured by shovels or boxes of sand?
23. Was the mortar of masonry cement or portland cement and lime?
24. Was the weather hot, warm, good, cool, or cold?
25. How many days were lost due to rain? Cold weather?
26. How far was mortar strung out on the bed joint before the brick was laid?
27. Were brick wet before laying? How often? How?
28. Did the brick tend to float on the bed joint?
29. Was the brick "hot," that is, highly absorptive?
30. What kind of wall ties and anchors were used?

31. If corrugated ties were used, could you bend them between your fingers?
32. Were inspectors ever around? Did you ever see the architect, engineer, or owner's project representative? How often?
33. Were test specimens ever taken of mortar from the mixer or off the board? How often? Were test prisms built? By whom? How often?
34. Were anchors turned down into the cores of brick? How were those cores filled?
35. Was any reinforced masonry used? If so, how and by whom was reinforcement placed?
36. Were any changes made in construction form plans or specifications?
37. Were nails or wooden line blocks used to hold the line?
38. Were expansion joints placed under shelf angles?
39. Were shelf angles placed before masonry was built or were they set on top of masonry?
40. Was the frame plumb and true or did brick have to be cut to pass or fit columns, spandrel beams, or shelf angles?
41. Was the flashing turned up at the ends, that is, end dams at control or expansion joints, columns, etc.? Were flashing joints sealed?
42. Were shelf angles mitered or were they cantilevered to be continuous around corners? Or were shelf angles discontinuous at wall corners?
43. What type of grout was used — slushed mortar, site mixed, or ready mixed?
44. What was the grout height?
45. Were cleanouts used? How many?
46. How was the grout consolidated?
47. Ask the respondent what he thinks caused the problem.
48. Has the respondent been questioned by anyone else about the job? Who? When?
49. Repeat question numbers 11, 16, 18, 27, 36, and 42.
50. Was the respondent generally cooperative and helpful?
51. Was the respondent generally specific and vague? Did he seem generally certain?

2.9 APPENDIX E

MASONRY TEST AND EXAMINATION METHODS

Material	Test or Examination	ASTM Method
1. Masonry Unit	Compression and Elasticity	C 67, C 140
2. Masonry Unit	Color	D 1729
3. Masonry Unit	Freeze-thaw	C 67, C 140
4. Masonry Unit	Water Absorption	C 67, C 140
5. Masonry Unit	Split Tensile	C 1006
6. Brick	Size, Warpage, and Chips	C 67, C 216, C 652
7. Mortar	Petrography	C 295
8. Prism	Compression and Elasticity	E 447
9. Prism	Flexure and Elasticity	E 518, C 1072
10. Prism	Bond Strength	C 952
11. Wall	Compression and Elasticity	E 72
12. Wall	Flexure and Elasticity	E 72
13. Wall	In-place Stress and Deformability	C 1196, C 1197
14. Wall	Shear and Rigidity	E 519
15. Wall	Water Permeance	E 514
16. Wall	Anchor Pullout	E 448, E 754

Chapter Three • *Evaluating Cracks*

The most frequent cause of masonry performance failure is not collapse but cracking. The design philosophy is to eliminate cracks or to limit crack openings to tolerable criteria. Cracks, however, can mar the architectural appearance, provide access for wind-driven rain, and indicate the potential for collapse. The structure's cracks are the first sign of distress and it is a warning that should not go unnoticed (Figure 3-1). Observation and evaluation are the key to diagnosing the cause of cracking and, if necessary, provide the framework to choose an appropriate repair technique.

3.1 *ARE ALL CRACKS DETRIMENTAL?*

The Portland Cement Association [1] suggests that the maximum crack width that will neither impair the surface appearance nor alarm the viewer is probably in the range of 0.010 to 0.015 inch, although wider crack widths may be tolerable. Unfortunately, there is no universal classification of detrimental crack widths since this depends on the structure's function and form. For a structure in which the architectural features were designed and constructed at a premium, a crack width of 0.010 inch may be totally unacceptable. A homeowner with a 20-year-old brick house would be very happy, and lucky, to have crack widths of only 0.010 inch.

Test results from Norwegian investigators [2] indicate that crack widths smaller than 0.004 inch will not allow wind-driven rain to enter. This does not necessarily imply that all crack widths greater than 0.004 inch will allow water penetration that is detrimental to the structure's durability, insulation resistance, or strength.

3.2 *CRACK CLASSIFICATION*

Many investigators [3] have proposed crack classification categories. A useful crack classification is shown below in Table 1.

TABLE 1. CRACK CLASSIFICATION [3]

Category		Crack Width, CW , inches
Very Fine	(Watertight)	CW < 0.004
Fine	(Exterior Exposure)	0.004 < CW < 0.008
Medium	(Interior Exposure - Wet)	0.008 < CW < 0.012
Extensive	(Interior Exposure - Dry)	0.012 < CW < 0.016
Severe		CW > 0.016

Figure 3-1. Observation and evaluation are the key to diagnosing the cause of cracking and, if necessary, provide the framework to choose an appropriate repair technique. This diagonal crack formed because of shear caused by lateral loads on the masonry structure. The crack is patched with plaster to determine if the crack is active (live or working) or non-working (dormant). If the crack is active, the patch will be cracked at a later date. Dormant cracks can be plugged with a rigid joint filler (epoxy or cement) but the active cracks can be filled only with a flexible sealant.

As shown, the crack classifications correlate with design crack widths for different exposures. Once the investigator classifies the crack, the exposure category also is determined. However, it's important to realize the expected variation in crack widths within a single member. For reinforced concrete structures, the coefficient of variation for crack widths of a single structural element is approximately 40% [4]. The coefficient of variation of crack widths in concrete masonry structural elements is likely to be of the same magnitude. Thus, the maximum crack width could be twice the average crack within a single member.

3.3 *CRACK OBSERVATION*

Crack observation denotes different activities to different people. Fortunately, it rarely means posting a 24-hour watch. Visual observation provides the first clues for the cause of cracking but only if properly documented. The investigator should record, with photographs and sketches, the following crack information:

- Pattern (horizontal, vertical, diagonal — straight or stepped)
- Length
- Width (uniform or tapered; if tapered note how)
- Depth (through paint, plaster, wall)
- Age (clean crack indicates new; coated with paint or dirt indicates old crack)

For crack width measurement, most investigators use a crack comparator (visual crack gauge) although more accurate measurements are possible with a graduated magnifying device (Figure 3-2). For long-term measurements, linear variable differential transformer (LVDT's) with automatic recording equipment are commercially available. The crack comparator is easy to use, sufficiently accurate for most jobs, and usually given away free by companies specializing in failure investigations. The graduated magnifiers cost between $50 and $100.

3.3.1 *Crack movement*

It is important to recognize the difference between an active or working crack and a dormant crack. An active crack is subject to further movement and a dormant crack is unlikely to get wider, close, or extend in length. A dormant crack can be created by a temporary overload, and a live (working crack) may be created by temperature changes that continually cause the crack to open and close. Also, a crack caused by one mechanism (overload) may turn into a working crack because of another mechanism (temperature changes).

Three relatively inexpensive methods are available to detect crack movements. The easiest is to measure the crack width with a crack comparator at regular time intervals, every day or week. Always measure the crack width at the same location. Draw a line across the crack to mark the measuring location. Do this at three or four places along the crack. If rain

Figure 3-2. For crack width measurement, most investigators use a crack comparator although more accurate measurements are possible with a graduated magnifying device. The crack comparator is easy to use, sufficiently accurate for most jobs, and usually given away free by companies specializing in failure investigations. The graduated magnifiers cost between $50 and $100.

or moisture are possible, use a permanent marker to indicate the crack measurement location. Record both crack width and date of recording.

Another inexpensive method is to mix a small amount of plaster to spot patch the crack. Use hot water to speed the setting time of the patch. Record the date the patch was placed. Inspect the patch at regular intervals to see if the patch has cracked. A cracked patch is an indication of a working crack. A word of caution: make sure the material used for the spot patch will not crack from drying shrinkage. Use a non-shrink patching material.

A two-piece crack monitor that sells for less than $15 can be used to observe the crack movement (Figure 3-3). The two-piece unit is connected by a piece of tape that holds the red cross-hairs at the zero mark. One plastic piece has the red cross-hairs and the other plastic piece has a grid system that measures the distance from the original zero mark. The monitor is attached across the crack so that each plastic piece can move independently. During crack movement, the red cross-hairs slide over the grid system indicating the amount of horizontal and vertical crack movement.

Attach the monitor to the wall with epoxy or a fast-setting glue. Allow the adhesive to cure (epoxy about 24 hours, fast-setting glue about 15 minutes) then cut the tape holding the two pieces together. If either side of the crack moves, the monitor moves. The red cross-hairs, originally at zero, can be read to determine the range of movement. It is usually preferable to read and record the cross-hair movement at the same time each day.

If the crack is dormant, then it can be fixed with a rigid filler (epoxy or cement). If the crack is live (or working) then flexible sealants should be considered as an appropriate fix.

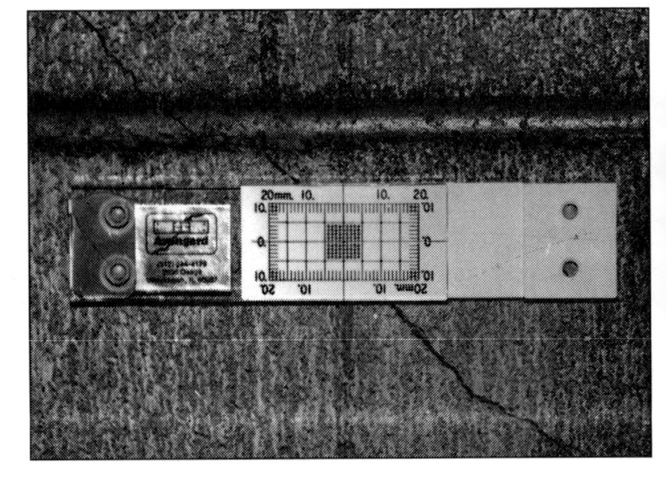

Figure 3-3. A two-piece crack monitor that sells for less than $15 can be used to observe the crack movement. The monitor is attached across the crack so that each plastic piece can move independently. During crack movement, the red cross-hairs slide over the grid system indicating the amount of horizontal and vertical crack movement. Attach the monitor to the wall with epoxy or a fast-setting glue. Allow the adhesive to cure (epoxy about 24 hours, fast-setting glue about 15 minutes) then cut the tape holding the two pieces together. If either side of the crack moves, the monitor moves. The red cross-hairs, originally at zero, can be read to determine the range of movement. It is preferable to read and record the cross-hair movement at the same time each day.

3.3.2 *Long-term crack widths*

It is difficult to estimate the increase in crack width caused by long-term or repetitive loading. For reinforced concrete, ACI Committee 224 [4] indicates that a doubling of crack width can be expected with time. A joint research program [5] by the National Concrete Masonry Association (NCMA) and the Portland Cement Association (PCA) concludes that the deformations of reinforced grouted concrete masonry walls are similar in magnitude to deformations of cast-in-place concrete. Therefore, for concrete masonry an initial crack width could double in size.

Clay masonry exhibits little creep [6] except in the mortar joints. Long-term deformation studies on brick piers suggest that crack widths in clay masonry are not likely to double under sustained or repetitive loading.

3.4 *WHY DO CRACKS OCCUR?*

Cracks result from strain that induces stress in excess of the material's capacity [3]. Strains can be caused by loads or by restraint of volume changes. Volume changes are created by changes in temperature and moisture and also can be induced by salt crystallization and corrosion. Loads can be caused by the dead weight of materials, the live load of equipment and people, soil and water pressure, wind, snow, and earthquakes. Deflections and deformations imposed by foundations, frames, roof slabs, and other structural elements also can induce cracking.

3.5 *CRACK EVALUATION*

The information obtained during the crack observation process will help guide the investigator in evaluating the cause of cracking. As shown previously, cracks can occur for many reasons and determining the cause of cracking requires good engineering judgment and experience. To assist investigators in their evaluation, Figures 3-4 and 3-5 illustrate typical crack patterns in block and brick masonry and their causes. The crack patterns and causes are based on References 1, 3, and 7.

3.6 *TROUBLESHOOTING CRACKED MASONRY BY LOCATION AND TYPE OF CRACK [9]*

3.6.1 *Location: Structural Frame Infill Walls*

Type of Crack: Constant-width stepped diagonal cracks originating in lower corners.
Cause: Top and bottom beams have deflected an equal distance under load, forcing masonry downward in the center of the wall.

Type of Crack: Vertical and diagonal cracks at bottom of walls, widest at the bottom of the crack.
Cause: Top beam has deflected more than bottom beam, compressing masonry.

Type of Crack: Vertical cracks at center of wall, with crack widest at its midpoint.
Cause: Top and bottom of wall are restrained from moving vertically.

Figure 3-4 [From Ref. 8]. Vertical deflection of concrete beam.

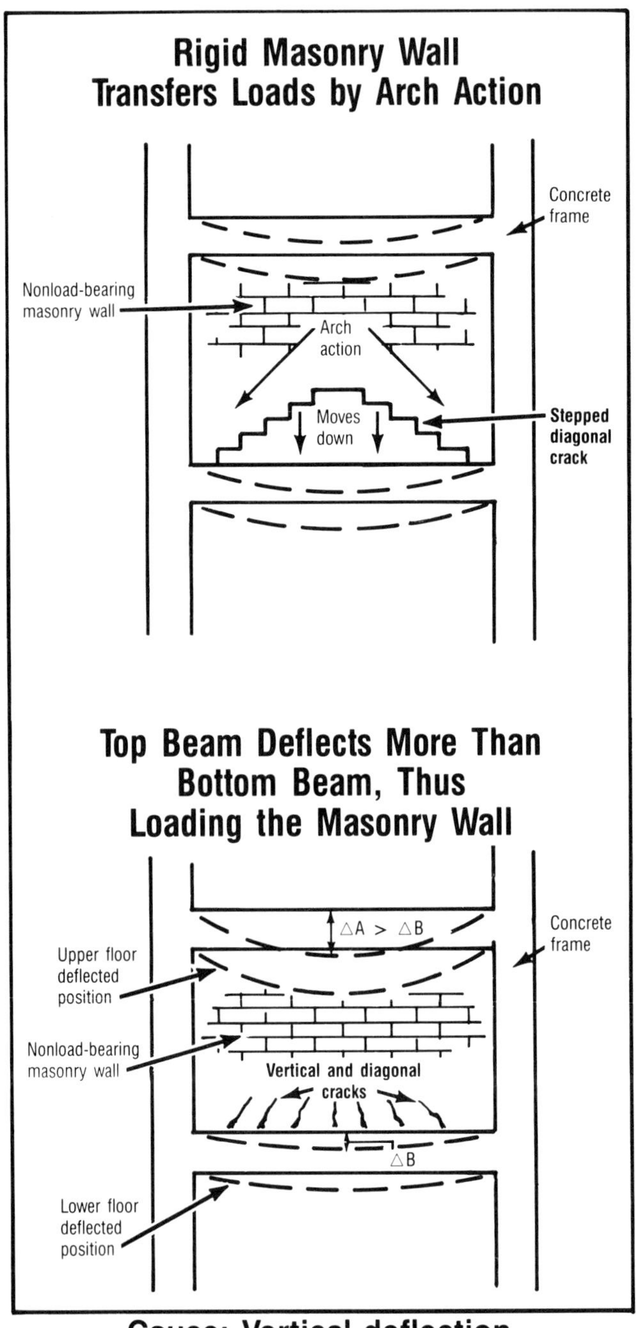

Figure 3-5 [From Ref. 8]. Foundation settlement or soil heave.

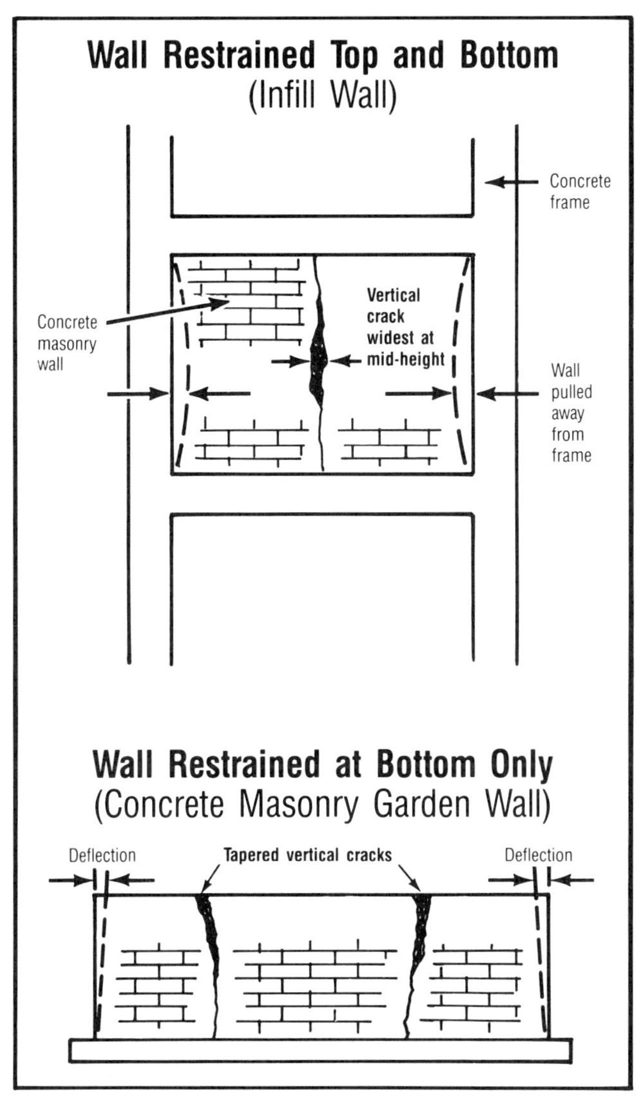

Cause: Shrinkage of mortar and concrete masonry units

3.6.2 *Location: Shelf Angles*

Type of Crack: Bowed masonry with horizontal and vertical cracks near shelf angle.

Cause: There's no room for the angles to expand because the ends have been butted together. When expansion occurs, restrained angles bow outward, upward, or both.

Type of Crack: Constant-width horizontal cracks and crushed masonry.

Cause: A. Without horizontal expansion joints, loads on frame are transferred to masonry, concentrating at shelf angles and window heads. Horizontal soft joints under the angles are needed to transfer the weight of the masonry to the structural system.

B. Missing and insufficient soft joints that have been mortared do not permit the load to be transferred by shelf angles to the structure.

C. Shelf angles rotate when not securely attached to the frame or not fully shimmed around anchor bolts. Rotation exerts pressure on masonry below and inadequately supports masonry above.

D. Steel angles oxidize, or rust, when exposed to moisture and the atmosphere, causing them to expand. This expansion, or exfoliation, causes a rusting lintel to act like a pry bar, forcing cracks in masonry.

Type of Crack: Constant-width vertical cracks at corner locations.

Cause: When shelf angles have been cut square on the ends at corner locations, there is an open area that provides insufficient support for masonry. To avoid this, miter or cantilever the shelf angles.

3.6.3 *Location: Near Lintels*

Type of Crack: Constant-width vertical and horizontal cracks at ends of lintel.

Cause: When ends of lintels 4 to 6 feet or more in length are installed tight against adjacent masonry, there is no room for the lintels to expand. To prevent this, leave a space between the ends of the lintels and adjoining masonry.

3.6.4 *Location: Mortar Joints*

Type of Crack: Constant-width vertical and horizontal cracks in wide mortar joints.

Cause: To make wider joints, masons usually use a mortar that contains less water and less lime, resulting in a greater proportion of cement. High-cement mortar undergoes more drying shrinkage and thus more shrinkage occurs.

Type of Crack: Constant-width vertical and horizontal cracks.

Cause: Caused by differential movement between masonry units and mortar. For example, if thermal expansion characteristics of masonry units are greater than that of the mortar, cracks may appear.

Type of Crack: Vertical and horizontal cracks.

Cause: Continuous joint reinforcement across movement joints may buckle when adjacent wall sections move, causing mortar to crack and to be pushed out.

Type of Crack: Various types of cracks.
Cause: Caused by disintegration of bond due to severe chemical or acid cleaning of masonry.

3.6.5 *Location: Spalled Brick*

Type of Crack: Constant-width vertical and horizontal cracks on old or historic masonry that has been repaired.
Cause: Using a higher-strength repair mortar on old and historic masonry can cause cracking because the mortar is too hard and doesn't allow individual brick to absorb slight movements caused by temperature changes, settlement, and vibration. For this reason, repointed mortar should be somewhat weaker than the masonry units and match the existing mortar color and strength as closely as possible.

Type of Crack: Constant-width vertical and horizontal cracks on spalled brick in buildings in freeze-thaw areas.
Cause: Water increases in volume by about 9% when it freezes. This expansion causes cracks when there is no room within brick for freezing water to expand. To allow for expansion of freezing water, brick that meets the guidelines recommended by ASTM C 216, Standard Specification for Facing Brick, should be used.

Type of Crack: Constant-width cracks and spalled brick on buildings in wet areas.
Cause: Recrystallization of soluble salts within masonry create great expansive pressures. This is especially true of buildings that have had water-resistant coatings applied. Although the coatings allow water to pass through as vapor, the salts remain behind.

3.6.6 *Location: Block*

Type of Crack: Constant-width cracks in block and mortar joints.
Cause: Because concrete block shrink as they dry, allowing them to remain exposed in inclement weather during construction or increasing their moisture content by wetting or soaking them before laying increases drying shrinkage, leading to cracking.

3.6.7 *Location: At Various Locations*

Type of Crack: Various-width cracks in masonry units and mortar joints.
Cause: Water flowing through walls can lead to differential settlement and deterioration of adjacent materials, causing masonry to crack. This can be especially prevalent in wood-frame buildings.

3.6.8 *Location: Expansion Joints*

Type of Crack: Constant-width straight or stepped cracks at corners or stepped cracks adjacent to openings.
Cause: Because walls expand in the direction of corners, too few expansion joints can cause corner brick to rotate and crack. This typically occurs near the first head joints closest to the corner.

Type of Crack: Constant-width vertical cracks between expansion joints.
Cause: Using too few expansion joints does not leave enough room for expansion. Expanding masonry can force sealant material out of expansion joints and create cracks.

Type of Crack: Constant-width vertical cracks near expansion joints.
Cause: Mortar bridging across movement joints defeats the purpose of the joints by preventing the joints from accommodating movement.

3.6.9 *Location: Walls Built on Foundations and Slabs*

Type of Crack: Constant-width horizontal cracks near bottoms of walls.
Cause: Brick walls on concrete foundations expand and the concrete foundation shrinks. This differential movement causes cracking when the wall is bonded to the foundation.

Type of Crack: Tapered vertical cracks, widest at the top and originating at the bottom of a wall.
Cause: Both corners of the foundation have settled or soil has heaved near center of wall.

Type of Crack: Stepped diagonal cracks at one corner of door openings or opposite corners of window openings, widest near opening.
Cause: The center of the foundation or slab has settled or soil has heaved at both ends of the wall.

Type of Crack: Stepped diagonal cracks at both corners of door openings, narrowest near the opening.
Cause: One end of a wall has settled.

3.6.10 *Location: Near Combinations of Dark- and Light-colored Masonry or of Different Materials in the Same Wall*

Type of Crack: Constant-width vertical and horizontal cracks.
Cause: Because dark-colored masonry absorbs more heat than light-colored masonry, it expands more. This difference in thermal expansion can cause cracks in areas where dark- and light-colored masonry adjoin.

Type of Crack: Constant-width vertical and horizontal cracks.
Cause: Again, differences in thermal expansion can cause cracks. To avoid this in areas where block and brick are horizontally adjacent, place

vertical movement joints in the block at the midpoint between two vertical movement joints in the brick.

3.6.11 *Location: Near Trees*

Type of Crack: Various widths and types of cracks near foundations.
Cause: Foundation movement can be caused by tree roots. Thirsty trees can dry up clay soils, leading to foundation settlement.

3.6.12 *Location: Buildings Near New Construction*

Type of Crack: Various widths and types of cracks near foundations.
Cause: If a new excavation lowers the water table, soil consolidation results, causing foundation settlement on nearby buildings.

Type of Crack: Various types, widths, and locations of cracks.
Cause: Nearby pile driving, blasting, and similar shocks can crack masonry.

3.6.13 *Location: Buildings Near Industrial Areas*

Type of Crack: Various types, widths, and locations of cracks.
Cause: Cracks can result from mechanical impact caused by trucks, forklifts, falling trees, and from vibration caused by heavy road or rail traffic.

3.6.14 *Location: Parapet Walls*

Type of Crack: Constant-width vertical and horizontal cracks.
Cause: A. Because parapets are exposed on three sides to moisture and temperature extremes, they become hotter, colder, and wetter than the walls below them. This causes differences in absorption and thermal expansion, which causes cracks.
B. Parapets lack a dead load of masonry above them to help resist movement, making them more vulnerable to cracks than masonry below.

3.6.15 *Location: Wall Setbacks and Offsets*

Type of Crack: Constant-width vertical cracks.
Cause: Parallel walls expand toward an offset, causing the offset to rotate, producing vertical cracks.

3.6.16 *Location: No Specific Locations*

Type of Crack: Tapered vertical cracks that widen for no apparent reason.
Cause: Hairline cracks can widen over time due to cyclic movement and racheting action. When a crack develops, minute particles can fall into the crack and prevent masonry from returning to its original position. When the crack opens again, slightly larger particles fall in while other particles work themselves deeper into the crack. The crack widens and a minor crack can become a major problem.

3.7 REFERENCES

1. Portland Cement Association. 1982. *Building Movements and Joints*. Skokie, Ill.: Portland Cement Association.
2. Birkeland, O., and S.D. Severndsen. 1963. Norwegian Test Methods for Rain Penetration through Masonry Walls. Paper presented at ASTM symposium, Masonry Testing, STP No. 320.
3. Grimm, C.T. 1988. Masonry Cracks: A Review of Literature. *Masonry: Materials, Design, Construction, and Maintenance,* ASTM STP 992. Philadelphia: ASTM.
4. American Concrete Institute, Committee 224. 1989. Control of Cracking in Concrete Structures, Part 3. In *ACI Manual of Concrete Practice*. Detroit: ACI.
5. National Concrete Masonry Association. 1977. Strength and Time-dependent Deformations of Reinforced Concrete. NCMA-TEK 84. Herndon, Va.: National Concrete Masonry Association.
6. Sahlin, Sven. 1971. *Structural Masonry.* Englewood, New Jersey: Prentice-Hall.
7. Copeland, R.E. 1957. "Shrinkage and Temperature Stresses in Masonry," *Journal of the American Concrete Institute,* vol. 28, no. 8.
8. Suprenant, Bruce. 1990. Evaluating Cracks. *Aberdeen's Magazine of Masonry Construction.* Vol. 3, no. 2.
9. Koski, John. 1992. Troubleshooting Cracked Masonry. *Aberdeen's Magazine of Masonry Construction.* Vol. 5, no. 5.

Chapter Four • Strength Testing

4.1 INTRODUCTION

Evaluation of existing masonry structures usually involve two strength criteria: the ultimate magnitude and the uniformity or distribution of strength throughout the structure. Surface hardness, penetration, or ultrasonic test methods can be used to estimate the variability of the strength within a structure. In many cases, however, nondestructive testing will not provide the information necessary for engineering analysis. Compressive, flexural, or shear strengths may be measured directly or estimated based on unit and mortar properties. The ultimate strength can be estimated by a variety of techniques, some of which are in-place and others requiring laboratory testing of specimens removed from the structure.

4.2 METHODS OF COMPRESSIVE STRENGTH TESTING

4.2.1 Prism Removal

Remove a section of the wall and send it to the laboratory for testing. This "prism" is tested according to ASTM E 447, Standard Test Methods for Compressive Strength of Masonry Prisms. Don't forget to apply the height-to-thickness correction factors for these specimens. These correction factors are shown in Table 1. This is one of the most common approaches to determining compressive strength, but it does have some problems. Removing samples from a wall affects its appearance, and repairing the wall might not be acceptable to the architect. Also, samples removed may be damaged during shipping.

4.2.2 Masonry Unit and Mortar Analysis

Another method requires removing samples of the masonry units and the mortar. The masonry units are tested according to ASTM C 67 (clay masonry) or ASTM C 140 (concrete masonry) to determine the unit's compressive strength. Mortar samples are chemically analyzed to determine the mortar type. Knowing the mortar type and the masonry unit

TABLE 1. CORRECTION FACTORS FOR STRENGTH*

	Prism Height-to-Thickness Ratio						
Masonry Type	2.0	2.5	3.0	3.5	4.0	4.5	5.0
Clay	0.82	0.85	0.88	0.91	0.94	0.97	1.0
Concrete	1.00	1.04	1.07	1.11	1.15	1.18	1.22

* From Specifications for Masonry Structures (ACI 530.1-92/ASCE 6-92/TMS 602-92).

strength, the compressive strength can be estimated. Tables 2 and 3 (from Specifications for Masonry Structures (ACI 530.1-92/ASCE 6-92/TMS 602-92)) provide an estimate of the compressive strength of the masonry based on the compressive strength of the masonry unit and the type of mortar. Note that these tables provide values for specified compressive strength f'_m to be used for design purposes and do not reflect actual compressive strengths as would be measured by in-place or laboratory testing.

TABLE 2. COMPRESSIVE STRENGTH OF CLAY MASONRY

Net Area Compressive Strength of Clay Masonry Units, psi		Net Area Compressive Strength of Masonry
Type M or S Mortar	**Type N Mortar**	**(psi)**
2400	3000	1000
4400	5500	1500
6400	8000	2000
8400	10,500	2500
10,400	13,000	3000
12,400	----	3500
14,400	----	4000

*From Specifications for Masonry Structures (ACI 530.1-92/ASCE 6-92/TMS 602-92).

4.2.3 *In-place Hydraulic Ram*

A prism test can be conducted in-place by isolating a small section of masonry and loading with a hydraulic ram [1]. Make two vertical cuts entirely through the wall at the proposed test area to isolate a masonry prism. Remove the masonry units between the vertical cuts above the test prism and insert a hydraulic ram or series of rams into the space. Cap the prism bearing surface using gypsum cement and use bearing plates to distribute the load. Be careful not to let the capping material penetrate into voids within the prism to be tested. Pressurize the ram to load the masonry in compression. This test will damage the wall and may be difficult to perform in multiwythe walls.

4.2.4 *Flatjack Testing*

Flatjack tests also are used for in-place compressive testing. The flatjack, a thin steel bladder, is inserted into a mortar joint slot and expands when pressurized, applying a compressive stress to the masonry. Some damage to the wall may occur but it's usually not difficult to repair. Flatjack testing is described in detail in chapter 9.

TABLE 3. COMPRESSIVE STRENGTH OF CONCRETE MASONRY

Net Area Compressive Strength of Concrete Masonry Units, psi		Net Area Compressive Strength of Masonry
Type M or S Mortar	**Type N Mortar**	**(psi)**
1250	1300	1000
1900	2150	1500
2800	3050	2000
3750	4050	2500
4800	5250	3000

* For units less than 4 inches in height, use 85% of the values listed.
* From Specifications for Masonry Structures (ACI 530.1-92/ASCE 6-92/TMS 602-92).

4.3 *REMOVAL OF TEST SPECIMENS FROM EXISTING STRUCTURES*

It is often desirable to remove specimens from existing masonry buildings for laboratory tests. Standard laboratory load tests can be used to determine masonry material properties such as compression, shear, and flexural strength; results from these tests can be used for analysis or correlated to nondestructive test results. Special care must be taken during specimen removal and preparation to obtain an undamaged sample that is representative of in-place masonry.

Specimens removed from existing masonry structures are tested in the laboratory using standardized tests to determine material properties. Testing of removed specimens can be conducted in controlled laboratory conditions to determine compressive strength and modulus, tensile strength, flexural bond strength, and shear strength. Established ASTM test methods may be used for many types of load tests, as shown in Figure 4-1. A listing of ASTM standards for masonry testing is provided at the end of this chapter.

Nondestructive and in-place tests will not always provide complete information for characterization of building integrity and condition. Information obtained by laboratory testing can be used for engineering analysis and also to determine correlations between nondestructive tests and material properties.

Removing specimens from existing masonry is not allowed in all situations. Building owners may not permit specimens to be removed from historically significant or critical structures. Disfigurement to the structure's appearance and effects on structural integrity necessitate repairs to damaged areas. Specimen removal and transportation is a difficult, expensive, and time-consuming prospect. The prohibitive cost of specimen removal, laboratory testing, and subsequent repairs limit the number of tests that can be conducted. A certain percentage of specimens can be expected to be damaged during the process of removal, transportation, and preparation for testing.

Several references address the issue of specimen removal. The only standardized method is provided by the European agency RILEM. RILEM LUM.D.1 [2] provides guidelines for "Removal and Testing of Specimens from Existing Structures." References 3 and 4 provide additional information on specialized fixtures and methods for specimen removal, capping, and preparation for testing.

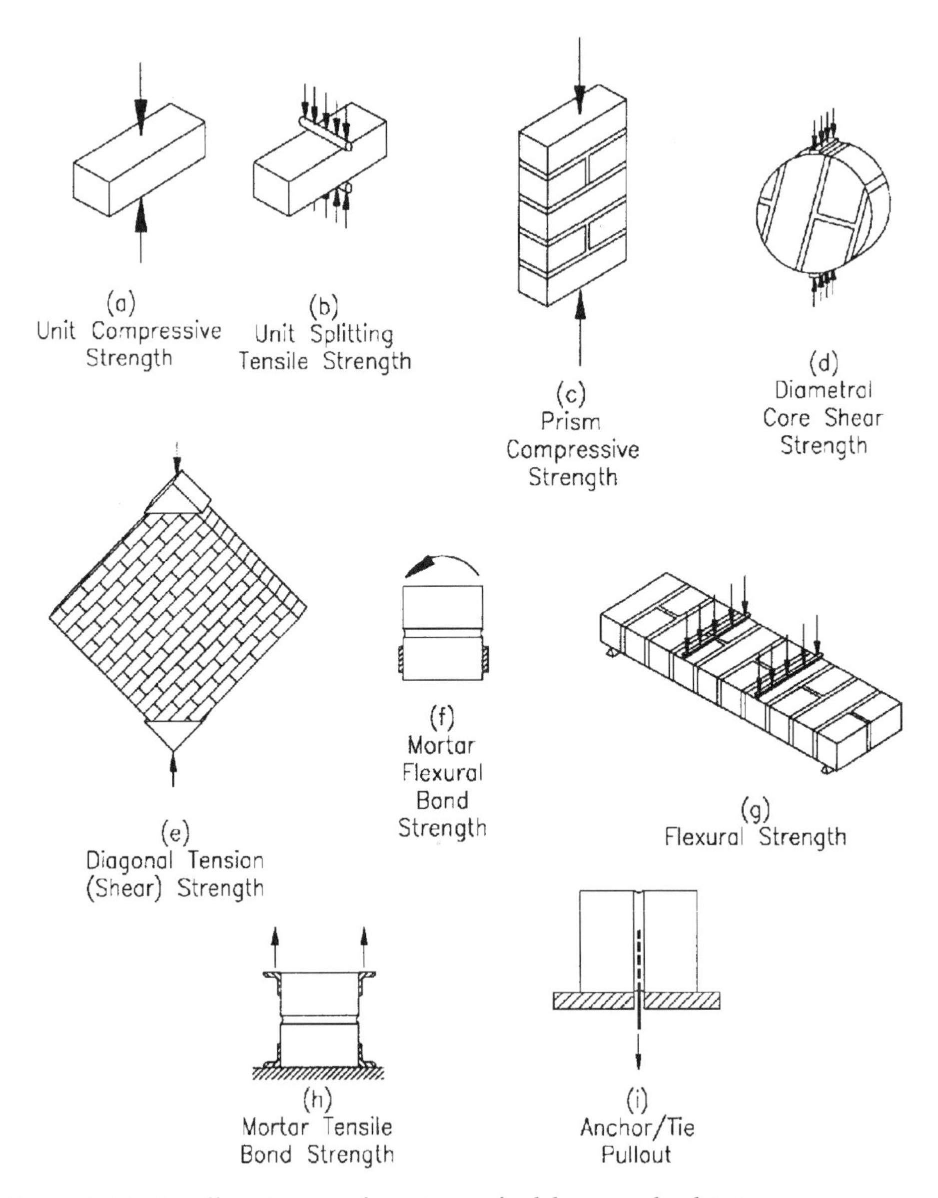

Figure 4-1 (a-i) Different types of specimens for laboratory load tests.

4.3.1 *Equipment*

Standard power and hand tools are used for removal of small mortar and unit specimens. Large saws and handling equipment are necessary when larger masonry samples are to be removed. Basic equipment for masonry specimen removal is listed below.

Removal of brick units and mortar specimens can be accomplished by using a rotary power drill with a masonry bit to remove mortar by stitch drilling. Use only rotary-type drills; hammer drills and power chisels disturb the specimen and should not be allowed.

Core drills equipped with hollow diamond-tipped bits are used to obtain cores for testing. Before drilling, core drills must be firmly attached to the surface of masonry walls using anchor bolts. Cores may range in size from less than one-inch diameter for small brick or mortar samples, to six or eight inches in diameter for diametral shear tests.

Removal of larger specimens require cutting of the masonry with a masonry or concrete saw. Diamond-tipped blades are recommended; abrasive blades can build up excessive heat and should not be used. If possible the blade should be able to cut through the entire wall thickness in a single pass. This requires using a saw with a large-diameter blade for cutting multiwythe masonry. Use a track-mounted saw for most cuts; a handheld saw is acceptable for shallow cuts and clearing mortar from joints.

Several manufacturers have developed innovative cutting equipment that can be used for masonry. Edge-drive saws use specially designed circular blades and have the main driving mechanism located at the edge of the blade. Deeper cuts are possible as a result of this configuration, with cuts up to about 80% of the blade diameter being possible. Special chain saws equipped with segmented diamond blades are useful for cutting through thick walls and eliminate overcuts necessary with circular saws. Diamond-impregnated wire saws are normally used for cutting large stones but also have been used successfully for removing specimens from massive concrete and masonry structures.

For most applications it is preferable to use water-cooled core bits and saws for specimen removal. A source of water and appropriate hoses is required.

Hand tools such as hammer, chisel, scraper, and wedges are useful for removing mortar and preparing the specimen for removal. Large clamps or fixtures are often necessary for confining large specimens during and after cutting. Clamping fixtures reduce damage to the specimen resulting from movement and transportation. Use a forklift or small crane for removing larger specimens. Test specimens must be firmly secured in a wooden frame or packing crate before transportation to the testing laboratory. Place packing materials around the specimen to prevent damage during transportation and handling.

Use appropriate safety equipment when cutting, sawing, and drilling. This includes eye and face protection, gloves, and protective clothing. Shoring is required when removing large specimens to maintain structural integrity.

4.4 REMOVAL PROCEDURE

The size of the test specimen is normally dictated by standards governing each specific test; general guidelines for compression, shear, and flexure specimens are included here. Whenever possible the specimen should be taken from the entire wall thickness. The minimum thickness for most tests is one wythe except when tests are intended for determination of composite action between multiple wythes or masonry veneer.

4.4.1 Specimen Sizes

Compression Specimens — Specimens for compression testing should have a minimum width equal to one unit, or the thickness of the specimen, whichever is greater (Figure 4-1 (c)). The height should be 12 inches or three courses (minimum), with a height/thickness ratio greater than or equal to 3. Specimens with a height/thickness ratio approaching 5 will be more representative of in-place masonry behavior [5].

Core Tests — Testing of cores for masonry shear strength is allowed in some jurisdictions. Cores for such tests should be 8 inches in diameter. The intersection of a bed and head joint should coincide with the center of the core as shown in Figure 4-1 (d).

Shear Specimens — The diagonal tension (shear) test described in ASTM E 519 uses a specimen that measures 4 feet on each side (Figure 4-1 (e)). Removal of such a large specimen is difficult and damaging to the structure. Tests on smaller specimens are possible but should be used with caution. Insufficient test data exists to correlate shear strength of smaller specimens with masonry shear behavior. Minimum size specimens should have a height greater than 12 inches or three courses of masonry. The width should be at least two units and must be equal to the specimen height. Specimens should be cut such that whole units are located at the loaded corners.

Flexural Specimens — Two types of flexural tests may be conducted, as shown in Figure 4-1 (f) and (g). Specimens for bond wrench testing (ASTM C 1072) should consist of one wythe of masonry and should have a minimum height of two units (Figure 4-1 (f)). A tall specimen with several joints for testing is most useful. Flexural beam specimens (ASTM E 518) also should be a single wythe thick and a single unit wide but must be a minimum of four units in length with a span to depth ratio of 2.5 or more (Figure 4-1 (g)).

4.4.2 Test Locations

Locations for specimen removal must be representative of variations in material quality and condition throughout the structure. Representative areas are chosen based on condition surveys and test samples should be removed from each area based on random sampling techniques. The actual number of test specimens is specified by applicable test standards and varies depending on the expected variation in test results, desired confidence level, and allowable statistical error.

In general it is easiest to remove the largest size sample possible, from which several test specimens can be taken. Large samples are transported to the laboratory for careful cutting into test specimens under controlled conditions. Size normally is governed by removal and transportation equipment and structural considerations. It is necessary to remove the entire wall thickness for multiwythe construction. If necessary for testing of a single wythe, multiwythe specimens can be cut along the collar joint in the laboratory.

Removal of single units and small mortar samples from sound masonry should be possible without special precautions, being careful to choose areas away from major load paths, arch lines, and slender walls or columns. It is important that structural integrity is not compromised. Employ the services of an engineer before and during removal of larger specimens. Shoring is necessary during removal of larger specimens or specimens from critical areas.

4.4.3 *Equipment Setup*

Carefully mark the outline of the specimen to be removed. Use a large-diameter saw whenever possible to cut through the entire wall thickness at once or drill pilot holes at the corners to align equipment on opposite sides of the wall. Attach guide angles to the wall to ensure straight cuts. It is important to obtain a specimen with parallel sides and straight cuts. Time spent for accurate setup of cutting equipment will save on laboratory trimming and capping later. Install shoring several courses above the cutout area, if necessary, to carry structural loads.

4.4.4 *Core Specimens*

Small-diameter cores may be tested to characterize unit properties and for interior inspection; larger-diameter 8-inch cores are tested for masonry tensile and shear properties. Use a truck-mounted core drill or attach the drill to the wall with anchor bolts to stabilize during coring. Use a steady, even drilling pressure to minimize damage to the sample. Avoid reinforcing bars when taking cores in reinforced masonry. Carefully remove cores from the wall and immediately place them in packing crates to avoid damage. Older masonry with weak mortar may not have sufficient bond to allow removal of intact specimens.

4.4.5 *Stitch-Drilling*

Small specimens, single units, and mortar samples are best obtained by using a rotary drill to remove mortar. Remove mortar from around sample units by stitch-drilling — successive drilling of small-diameter holes (approximately equal to the joint thickness) in the mortar joint. Remove mortar between holes using the drill or a hand chisel and a scraping motion. Avoid hammer blows and prying motions, which have a tendency to damage specimens. The mortar in the joint below the unit should be fully removed as the first step; the unit will normally fall under its own weight during removal of mortar from the head joints and top joint.

4.4.6 *Large Specimens*

To avoid damaging mortar-unit bond keep the sample in compression normal to bed joints during cutting by using wedges, clamps, or specially designed frames. Removal of large specimens can be difficult and careful planning is essential for a successful operation. First, take out several units entirely through the wall along the bottom surface of the specimen. Install lifting apparatus such as the forks of a forklift or cribbing for external support through the cleared space below the specimen. Provide full base support with a heavy steel or timber member. Follow this step by cutting along the top of the specimen and then the sides. Take out a layer of brickwork from the specimen's boundary to provide clearance for easy removal. A support frame or other clamping apparatus should be built around the specimen in-place to prevent damage during removal. Lift carefully using a crane or forklift and place into the transportation vehicle.

4.4.7 *Transportation*

Specimens must be packaged securely for transportation using a crate or wooden framework and cushioning materials. Keep large specimens in compression normal to bed joints using a frame and clamps as shown in Figure 4-2. Transport the specimen in an upright position, brace it against overturning, and tie it to the vehicle to prevent movement.

4.4.8 *Storage*

During transportation and before testing, the specimen should be stored in conditions similar to those experienced in the structure. If the specimen was wet or saturated in-place, be careful not to let it dry as this may result in formation of shrinkage cracks and debonding of mortar from units. If the specimen was wetted significantly during cutting or coring, it must be allowed to dry before testing. Dry slowly under controlled situations to prevent crack development.

4.4.9 *Cleanup and Repairs*

Whenever possible save all units removed from the masonry for post-test repairs. Otherwise repair the masonry with units of similar color and material properties. Mortar for repairs should be similar in composition and color to the original mortar. Where cores have been taken, it will be necessary to remove units damaged by the core drill to provide an easily repaired section. Shore the area adequately until the repairs have gained sufficient strength to withstand any structural loads.

4.5 *PREPARATION FOR TESTING*

4.5.1 *Trimming*

If necessary, the sample may be trimmed and portioned into appropriately sized test specimens. This is best accomplished in controlled laboratory conditions using track-mounted saws. Do not test any masonry that was damaged during removal and transportation. Use ultrasonic or mechanical pulse velocity before removal and again in the laboratory to determine if the specimen has been damaged.

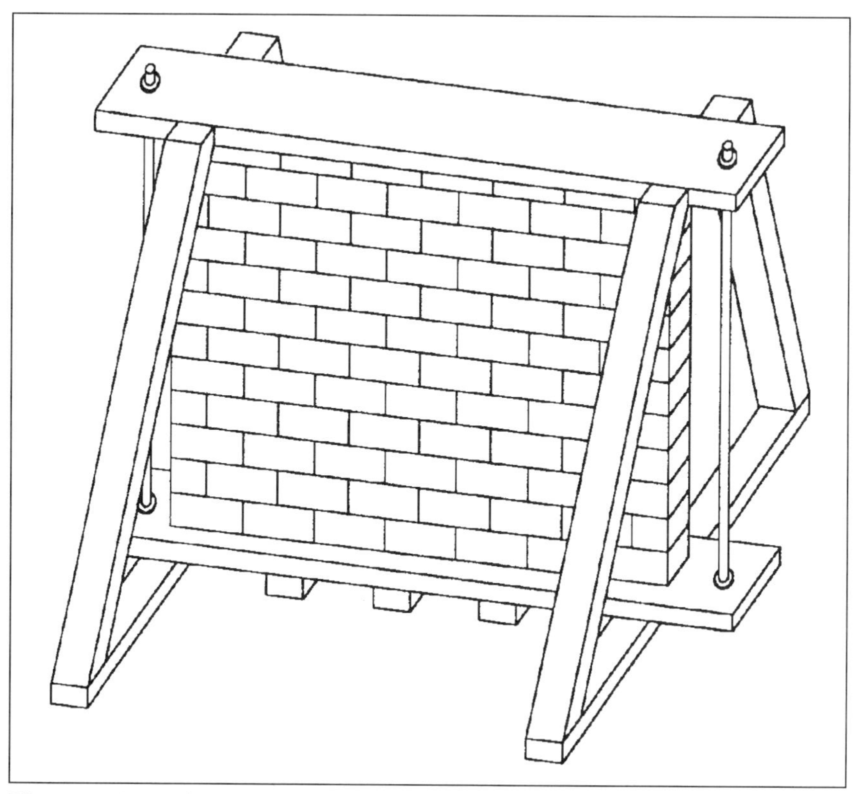

Figure 4-2 Confinement frame for transportation of large masonry specimens.

4.5.2 *Capping*

After the specimen has been cut to the correct size and trimmed to remove irregular edges, loading surfaces must be capped (see Ref. 3, 5, and ASTM E 447 for capping of compression prisms). Use a fluid mixture of gypsum cement to fill small surface irregularities and provide smooth, parallel bearing surfaces. Mix the cement in a small container according to the manufacturer's recommendations. Vibrate or tap the container to remove entrapped air after mixing. Pour the mixture onto a lightly oiled flat metal or glass surface. While lifting the specimen do not subject mortar bed joints to tensile stresses. Lift the specimen from the bottom unit to prevent separation of bed joints. Lower the specimen into the gypsum cement in one smooth motion without rocking to prevent air from being trapped. Use a bubble or level to ensure the two caps are parallel or, alternatively, an alignment fixture may be used. An apparatus that clamps to the specimen for lifting by a forklift or overhead crane is useful.

After the gypsum cement has cured, rotate the specimen for capping on the other surface. If the specimen is large or fragile, it may not be feasible to rotate for capping of the top surface. In this situation, it is acceptable to construct a small frame around the top, into which the capping material may be poured. If the mixture is fluid, it will be self-leveling. Close any large voids or cells that may cause the interior to be filled with capping material. The average cap thickness must not exceed 0.125 inch.

4.5.3 *Labor Requirements*

Cores can be removed rapidly and will generally require from one to four man-hours per core, depending on size and type of masonry. Small unit specimens and assemblages also are fairly easy to remove from masonry structures. Older, weak mortars are easy to drill and cut; expect to take one to two hours for cutting and extraction of small prism specimens. Removal of specimens from modern masonry with cement-based mortars will require slightly more time. It could take 15 to 30 man-hours or more to remove a large (4 foot x 4 foot x 1 foot) specimen from a masonry structure and transport it to the laboratory. Finish cutting and capping require another one to four man-hours, depending on cutting required and size of specimen.

4.6 *STANDARDS FOR LABORATORY TESTING OF MASONRY UNITS AND ASSEMBLAGES*

Following is a listing of ASTM and Uniform Building Code (UBC) standards describing various techniques for determining masonry compressive, tensile, shear, flexural strength and water penetration properties. These standards have been developed for testing of laboratory or field-constructed specimens and do not specifically address testing of specimens removed from existing masonry buildings. These methods may be used by approval of the engineer or building official.

ASTM C 67-91, Standard Test Methods for Sampling and Testing Brick and Structural Clay Tile
ASTM C 140-91, Standard Methods of Sampling and Testing Concrete Masonry Units
ASTM C 901-85, Standard Specification for Prefabricated Masonry Panels
ASTM C 952-91, Standard Test Method for Bond Strength of Mortar to Masonry Units
ASTM C 1006-84, Standard Test Method for Splitting Tensile Strength of Masonry Units
ASTM C 1072-86, Standard Method for Measurement of Masonry

Flexural Bond Strength
ASTM E 447-84, Standard Test Methods for Compressive Strength of Masonry Prisms
ASTM E 448-90, Standard Test Methods for Strength of Anchors in Concrete and Masonry Elements
ASTM E 514-90, Standard Test Method for Water Penetration and Leakage through Masonry
ASTM E 518-80, Standard Test Methods for Flexural Bond Strength of Masonry
ASTM E 519-81, Standard Test Method for Diagonal Tension (Shear) in Masonry Assemblages
ASTM E 754-80, Standard Test Method for Pullout Resistance of Ties and Anchors Embedded in Masonry Mortar Joints
UBC Standard 21-9, Unburned Clay Masonry Units and Standard Methods of Sampling and Testing Unburned Clay Masonry Units
UBC Standard 21-20, Standard Test Method for Flexural Bond Strength of Mortar Cement

4.7 *LARGE-SCALE LOAD TESTS*

Portions of a masonry structure can be loaded in place to verify overall structural performance. Loads can be applied to simulate proposed live loads including wind loadings or seismic forces. Methods for determining applied loads and analyzing masonry structures are described in FEMA 178, NEHRP Handbook for the Seismic Evaluation of Existing Buildings [8], The ABK Methodology TR-08 [9], masonry design textbooks, and local building codes. Chapter 20 of ACI 318-89, Building Code Requirements for Reinforced Concrete [10], contains a detailed description of procedures for load testing.

Large-scale load tests are conducted for several reasons:

- Uncertainty of load paths and behavior of complicated structures, including two-way spanning, complex edge conditions, the effect of openings, and interaction of multiple wythes
- Verification of analytical models
- Inadequate information regarding archaic building materials and construction techniques
- Proof of structural adequacy under proposed load conditions
- Concern that material properties determined using localized methods may not be representative of overall behavior

For masonry, load tests are most often conducted on wall panels to investigate behavior when subjected to wind and seismic loads. Wall segments can be isolated and tested under either in-plane or out-of-plane loads to simulate wind and seismic forces. Floor and roof spans are tested to determine response to loadings and interaction with the masonry support structure.

Vertical loads on roof and floor spans are simulated by distribution of dead weights over the area in question. Loads are generated by placement of sand bags, bricks, or lead pigs at regular intervals throughout the area in question. Placement of dead loads is a time-consuming, labor-intensive process. Water placed in large tubs, pools, or barrels can also be used for simulation of vertical loads. The loading setup must be configured to prevent ponding as deflections occur.

Horizontal in-plane shear loads are applied using hydraulic cylinders. A description of large-scale in-place tests for determining masonry shear behavior is described in chapter 6 on shear testing. Out-of-plane wind and seismic forces also can be simulated using hydraulic equipment and point or line loads applied at various locations, reacting against a steel frame attached to the foundation, floor slab, or interior columns. An air bag or vacuum frame apparatus is more convenient for applying uniform loads and has been used successfully for out-of-plane field tests of masonry walls. This setup requires a reaction frame attached to the floor slab or self-equilibrating with the masonry wall. An internal pressure of 1 psi or less is sufficient for most prescribed loadings. Air bag loading also provides an inherently safer approach for preventing total collapse of the wall: large deflections preceding collapse increase the volume of the air bag and reduce the internal pressure, decreasing the overall applied load.

Figure 4-3 Lateral load test on an exterior wall. Uniformly applied wind loads were simulated using an air bag apparatus, shown here, between the wall and an external steel reaction frame.

Deflections are monitored closely during load application. Member strains and overall deflections are used for comparison with calculated deflections and for confirmation that serviceability criteria are met. A rapid increase in deflections or continuing deflection under constant load indicates that the safe load capacity has been surpassed. In such a situation, purge valves built into air-and-water loading systems permit rapid load reduction. Removal of dead weights require precious time in such situations and can be dangerous to personnel. Shoring must be placed to prevent sudden and catastrophic collapse while allowing expected deflections to occur.

4.7.1 *Simulated Wind Loading*

A clay-tile masonry structure constructed in 1942 was evaluated after 50 years of service before adding a new roof structure. Simplified analysis showed some wall and pilaster configurations were marginal for resisting design wind loads, and load tests were proposed for verification of structural adequacy. Two out-of-plane load tests were conducted on selected wall panels. Wind loads over an entire 14-foot square panel were simulated with an air bag and reaction frame (Figure 4-3). An internal air bag pressure of less than 0.2 psi was required to obtain the specified design wind load of 20 pounds per square foot. The wall was instrumented to track overall wall deflections, movement at the foundation and roof connections, and localized strains in critical areas. Subsequent loading proved the walls to be sufficiently strong to withstand design wind loadings without causing abnormal deflections or cracking. Some permanent minor offset (less than 0.060 inch) was recorded at the foundation and roof level.

4.8 REFERENCES

1. Chiostrini, S., and A.Vignoli. 1994. In Situ Determination of the Strength Properties of Masonry Walls by Destructive Shear and Compression Test. *Masonry International.* Vol. 7, no. 3.
2. RILEM. 1992. Removal and Testing of Specimens from Existing Structures, RILEM, LUM.D.1. Paris: The International Union of Testing and Researching Laboratories for Materials and Structures.
3. Fattal, S.G., and L.E. Cattaneo. 1977. Evaluation of Structural Properties of Masonry in Existing Buildings. *National Bureau of Standards Building Science Series.* No. 62.
4. Jones, W.D., and M.B. Butala. 1993. Procedures and Fixtures for Removing, Capping, Handling, and Testing Masonry Prisms and Flexural Bond Specimens. *Masonry Design and Construction, Problems and Repair,* ASTM STP 1180.
5. Kingsley, G.R., J.L. Noland, and M.P. Schuller. 1992. The Effect of Slenderness and End Restraint on the Behavior of Masonry Prisms. A Literature Review. *The Masonry Society Journal.* Vol. 10, no. 2.
6. Atkinson-Noland & Associates Inc. and the University of Colorado. 1985. *Preparation of Stack Bond Masonry Prisms.* Boulder, Colo.: Atkinson-Noland & Associates Inc. and the University of Colorado.
7. Pistone, B., and R. Roccati. 1988. Testing of Large Undistributed Samples of Old Masonry. Presented at 8th International Brick/Block Masonry Conference in Dublin, Ireland.
8. Federal Emergency Management Agency, FEMA 178. 1992. NEHRP Handbook for the Seismic Evaluation of Existing Buildings. Washington, D.C.: Building Seismic Safety Council.
9. ABK. 1984. *Methodology for Seismic Hazards in Existing Unreinforced Masonry Buildings,* Topical Report 08. Methodology prepared for the National Science Foundation, Contract No. NSF-C-PFR78-19200.
10. American Concrete Institute. 1992. *Building Code Requirements for Reinforced Concrete,* ACI 318-92. Detroit: ACI.

Chapter Five • Surface Techniques

5.1 REBOUND HAMMER

Measurement of masonry surface hardness is conducted using a specially designed impact hammer that measures the rebound of a mass from the tested surface to indicate masonry condition and unit material properties. The method is useful for preliminary nondestructive examinations for locating poorly constructed or deteriorated areas. Surface hardness readings can be taken in a rapid manner by ordinary personnel to effectively inspect large areas of masonry.

5.1.1 Background

A rebound hammer for determining surface hardness of concrete was developed in 1948 by Dr. Ernst Schmidt, a Swiss engineer. Since that time the Schmidt hammer has experienced widespread use for evaluating not only concrete, but timber, rock, and masonry as well. The basic concept of the hammer is quite simple: a mass is driven against the masonry surface by a spring-actuated force. On striking the surface the mass rebounds to provide an indication of material surface hardness. The rebound distance is greater for hard, dense materials than for soft materials and thus provides an indication of near-surface density, flaws, and deterioration. The test also provides a good indication of the elastic properties of masonry and has been correlated to compressive strength and the quality of mortar surrounding the test unit [1]. It is theorized that movement of the rigid masonry brick in a soft mortar foundation as it is impacted affects the rebound number.

Schmidt hammer tests are used for rapid, inexpensive investigations to determine variations in masonry condition throughout the structure. The method is most useful for determining surface hardness of masonry units and to indicate the general condition or quality of units, mortar, and assemblages. Test results are analyzed and used to delineate poorly constructed or deteriorated areas, which will require further testing using more advanced investigative techniques. Equipment for measurement of surface rebound hardness is inexpensive and can be operated by average personnel. The technique is especially useful for locating areas that have experienced surface damage by freeze-thaw action, salt crystallization, moisture penetration, or fire.

The technique is regarded as being nondestructive; however, soft units may be damaged slightly by the rebound hammer, which has a tendency to leave small pockmarks on the masonry surface. The primary usage of the method is for providing qualitative evaluations, with secondary usage for indication of material properties. Indication of masonry material properties requires correlation of rebound number to results of destructive tests for each particular combination of mortar and unit type being investigated.

The measured rebound number is affected by a number of variables, including surface roughness, specimen size, vicinity of the test to free edges, masonry water content, direction of test, and angular orientation

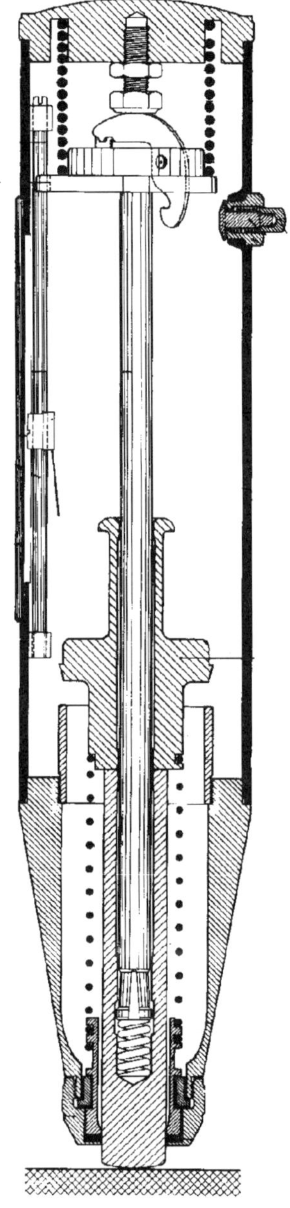

Figure 5-1. Impact hammer for measurement of masonry surface hardness.

to surface. Despite these limitations the technique does provide a fairly reliable method for nondestructive evaluation of masonry.

At the current time there are no standards for measuring surface hardness of masonry with the Schmidt hammer. The European agency RILEM is preparing a method for determining mortar surface hardness using a pendulum hammer [8]. Guidelines provided by ASTM C 805, *Test Method for Rebound Number of Hardened Concrete,* may be adopted for masonry evaluation. The International Society for Rock Mechanics (ISRM) provides an alternate method, *Suggested Method for Determination of the Schmidt Rebound Hardness,* which is more directly applicable toward masonry testing.

5.1.2 *Equipment*

The Schmidt hammer, shown in Figure 5-1, consists of a hardened steel plunger with a spring-actuated mass. The unit is placed against the masonry surface and the spring is released, driving the mass against the masonry. The mass rebounds from the surface and the rebound distance is read off an arbitrary scale.

Several types of hammers are available with different impact energies for different materials. The Type N hammer is useful for modern fired clay units; however, it may damage concrete masonry or older clay units. A Type L hammer has a lower impact energy for use with softer masonry. Mortar in joints is evaluated using a Type P hammer (Figure 5-2), a low-energy pendulum-type device specifically designed for low-strength materials.

Figure 5-2. Pendulum-type hammer for use on mortar and sensitive materials.

Normal surface hardness testing can be conducted by one person, but it is more efficient with two: one person conducts the test and the other records rebound numbers. Some hammer models have integral recording devices available. These devices record individual rebound numbers, on either a paper strip or digital device, for later analysis. Use of hammers with attached recorders is beneficial when a large number of readings are to be taken and require only a single person to conduct the test.

5.1.3 *Test Procedure*

Most surface hardness investigations will be conducted by laying out a gridwork and taking readings at regular intervals. It is important that test areas be chosen to capture the range of material variations throughout the structure.

The point of impact should be at least two units away from building edges, window openings, and doors and should be centered on the test unit. Cracked or loose units should be avoided. It will be necessary to smooth rough surfaces by grinding or rubbing with a carborundum stone.

Rebound hammers must be calibrated at regular intervals to guarantee repeatability of hardness readings. The calibration is obtained by impacting a hardened steel anvil (purchased from manufacturer) and used as a check of equipment dependability. The calibration device includes a tubular guide for ensuring the rebound hammer is perpendicular to the surface. Variations from the factory calibration number are taken into account by applying a correction formula to measured rebound numbers. Minor deviations from the specified calibration rebound number usually can be resolved by disassembling the unit and cleaning the guide rod.

Two different philosophies currently exist for the determination of rebound hardness. The ASTM method specifies taking 10 readings using a specified pattern, impacting the material surface only once at each test location. The RILEM technique is similar and uses nine measurements taken on mortar joints spaced uniformly over the test area. This type of test provides an indication of the surface condition of the unit or mortar being tested; however, it is highly affected by the near-surface microstructure and may not provide consistent results. ISRM describes an alternative technique [2] that uses multiple impacts at each point and is considered to provide a better indication of masonry compressive properties and condition of the unit-mortar matrix.

All tests are to be carried out with the hammer normal to the masonry surface. Most tests are conducted on vertical wall surfaces, with the hammer oriented in a horizontal position. Correction formulae are provided by the manufacturer to adjust readings from tests where the tested surface is a floor slab, ceiling, or arch. The tested surface must be smooth and free of dirt and loose material. A small abrasive stone is used to smooth rough surfaces.

The ASTM test requires 10 separate readings from single impacts in the chosen test area. Individual readings must be separated by at least 1 inch. Outliers are discarded and an average of the remaining readings is taken as the rebound hardness of the test area.

The ISRM test specifies that the test point be impacted three to four times, until the rebound number begins to stabilize. Ten successive readings are then recorded, without removing the plunger tip from the masonry. Because any errors in testing (such as slight angular misalignments) results in a reduction in rebound number, the five lowest readings are discarded. The rebound number for the test location is calculated as the mean of the five highest rebound numbers.

Surface hardness tests are quite fast and can be conducted by one or two technicians. Each reading will take only a few minutes, including

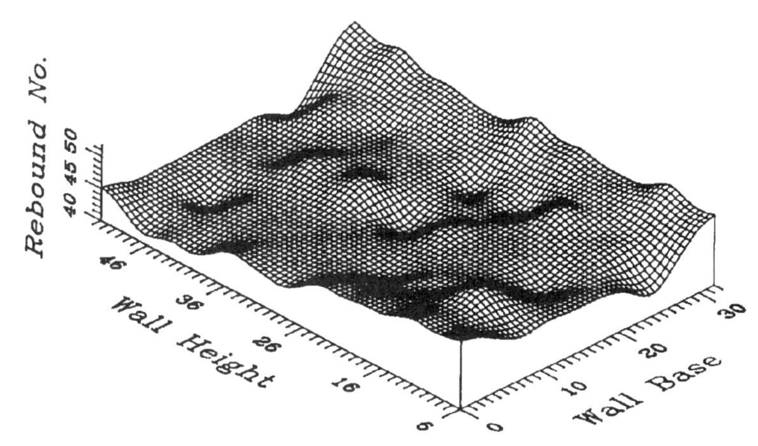

Figure 5-3. Surface plot of Schmidt hammer rebound number measured over the surface of a masonry wall.

locating, preparing, and measuring the test area, and determining and recording the rebound number. Hence a large area of masonry can be rapidly evaluated using the rebound hammer.

5.1.4 *Analysis/Interpretation of Results*

Schmidt rebound hardness tests are normally conducted as part of a condition survey or may be used to provide a preliminary indication of material property variations over large surface areas. A large number of rebound tests are conducted in each area, generally on a gridwork of about 8 to 16 inches. The readings are tabulated for analysis. Alternatively, readings taken on large wall surfaces may be used to form a contour or surface plot of rebound number (Figure 5-3). Contour plots are especially useful for determining general trends in surface hardness readings. The plot of rebound numbers shown in Figure 5-3 displays a deterioration in surface hardness toward the left edge of the wall. Readings significantly less than the mean value for the structure indicate deteriorated or deficient areas requiring further investigation using other more sophisticated techniques. Statistical tools such as the F-Test [3] may be used to determine the significance of rebound number variations.

The correlation between rebound number and masonry compressive strength is tenuous, at best. Laboratory investigations [1,4,5] have not provided sufficient evidence to either confirm or refute this application. If the main intent of the survey is to provide an indication of masonry compressive strength, destructive tests must be conducted on each combination of mortar and unit type present to correlate rebound number with compressive strength.

Research conducted on walls constructed using the same units but different types of mortar has shown that the test can be used as an indicator of mortar quality and compressive strength [1]. However, other investigations [4,5] have found little confidence in such correlations, concluding that the technique be limited to the evaluation of relative material uniformity and not be intended for determining masonry compressive strength.

5.1.5 *Summary of Rebound Hammer*

Masonry surface hardness measured using the Schmidt rebound hammer is useful for indicating general material condition. The method provides a simple, nondestructive technique for rapid preliminary condition surveys. Results from rebound hardness testing provide the basis for planning of more detailed investigations.

5.2 *PROBE PENETRATION*

Probe penetration tests are used extensively for indicating strength of in-place concrete. The technique is applicable to evaluation of masonry units and mortar and provides information on material property and condition. Steel probes or pins are driven into the masonry using a compressed spring or small explosive charge. The depth of penetration is correlated to material properties. Probe penetration tests are useful for preliminary condition evaluation of in-place masonry but are not totally nondestructive. Removal of embedded probes will leave small pockmarks and spalls on the masonry surface.

5.2.1 *Background*

Penetration testing uses specially designed equipment to provide a precise impacting force to the penetrating probe. A pneumatic force, gener-

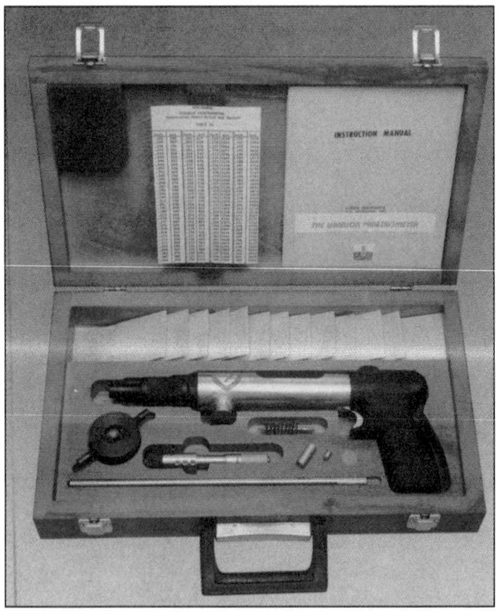

Figure 5-4. Powder-actuated probe penetration equipment.

ated by firing of a powder charge, drives a hardened steel probe into a masonry unit or mortar. An alternate type of penetration equipment uses a spring to drive a small-diameter pin into the masonry. The depth of penetration is measured with a micrometer to provide an indication of material properties. Probes penetrate further into weak, soft, or porous materials than harder, more dense materials. The method also can be used to quantify variations in masonry condition: probes will penetrate deeper into deteriorated masonry than sound masonry.

Equipment for probe penetration is commercially available, portable, easy to use (Figure 5-4), and provides immediate results for rapid field evaluations. Penetration tests can be used on all types of brick, block, and mortar.

A complex state of stress exists around the tip of a probe during impact. Depth of penetration is affected by strength, stiffness, and density of surrounding material and to some extent by elastic deformation of the mass into which the probe is fired. Penetration tests appear to be more useful for determining unit mortar strength alone instead of compressive strength of masonry prisms or assemblages. In addition, there is typically a significant amount of scatter in test results, requiring a large number of individual readings to provide adequate confidence in test results.

The force delivered to the probe or pin by the firing mechanism must be consistent from test to test. The driving energy generated by spring-actuated devices for small-diameter steel pins can be measured using calibration blocks provided by the manufacturer. There is no method for field calibration of the high-energy firing apparatus used to drive steel probes or to check the force delivered by powder charge.

Laboratory studies of penetration tests on masonry do not show consistently favorable correlations between probe penetration and compressive strength. Donovan [4] reports correlations between 0.61 and 0.94 between probe penetration into mortar joints and prism compressive strength. A lower degree of correlation was reported by Noland [1]: 0.62 between unit penetration and prism strength; 0.53 for mortar penetration and prism strength.

ASTM C 803, *Standard Test Method for Penetration Resistance of Hardened Concrete,* describes the technique for concrete evaluation. The method can be applied to masonry evaluation with minor modifications.

5.2.2 *Equipment*

Equipment for penetration testing is commercially manufactured. Equipment commonly used for concrete testing may be used for masonry; a variation of the standard probe system has been modified specifically for use with masonry. Required apparatus for masonry penetration testing is described below.

5.2.2.1 *Actuator*

Probe penetration equipment is portable and compact, consisting of a pistol-type apparatus that directs a powder-actuated charge toward the probe. A blank .22 caliber cartridge is inserted into the chamber, and a probe is inserted in the end of the cylinder. The device is actuated by pulling the trigger. A safety mechanism prevents firing unless the probe is pressed firmly against the masonry surface.

The device has two power settings: low power, for mortar and soft units with an expected compressive strength of less than 2000 psi, and a high power setting for hard clay units and grouted masonry.

Spring-loaded actuators also have been used for penetration testing. These devices have a lower energy output than powder-actuated drivers and are used to drive small-diameter steel pins.

5.2.2.2 *Probe*

The probe consists of a 0.25-inch-diameter hardened steel cylinder inside a plastic sleeve. Probes should be threaded on the exposed end to ease removal from the masonry. Steel pins for use with spring-actuated devices have a length of approximately 1.25 inches and a diameter of 0.15 inch.

5.2.2.3 *Penetration Measurement*

The depth of penetration is measured using a micrometer, accurate to 0.001 inch.

5.2.2.4 *Safety Equipment*

The probe penetration test uses a device similar in design and operation to a small firearm. Explosive charges can cause small spalls and shattering of the masonry surface. Spring-loaded devices also occasionally cause small chips to fly from around the pin. Full face and eye protection is required when operating the equipment.

5.2.3 *Test Procedure*

Test locations are chosen carefully to represent the normal variation in material properties and condition throughout the structure being investigated. Random locations are chosen in each general area to be investigated. Units that are cracked, loose, or otherwise damaged should not be tested. In addition, it is important not to test directly over hollow cells or interior cavities; tests must be carried out over the web or solid portions of masonry units. If mortar joints are to be tested, they must be completely filled as subsurface voids or furrows in the mortar will cause faulty readings. Struck joints are easier to test and provide more reliable results than tooled or raked joints. The test area must be relatively smooth; rough surfaces require grinding to provide a 1.5-inch-diameter bearing surface for the end of the penetrometer actuator.

The number of tests varies depending on the size of the region being investigated. ASTM C 803 recommends that for concrete the average penetration depth of three probes or six pins constitute one test for each specific area. Probe tests must be separated by at least 7 inches; pins are located 2 to 6 inches from adjacent pin penetration tests. Penetration tests can be conducted rapidly and require only about five to 10 minutes for each set of readings.

The device is pressed firmly against the masonry surface at the desired test location. The probe must be held perpendicular to the surface; this is checked using a guide rod provided by the manufacturer. The trigger is pulled, firing the explosive charge. The device is removed from the wall, leaving the embedded pin or probe and the sleeve around the probe.

Probes must remain firmly embedded in the masonry after firing. Tap lightly with a small hammer to identify loose probes; discard any probes that are inadequately seated. The depth of penetration is measured for each probe using a micrometer. Pins are removed from the masonry immediately after being driven, and the depth of the hole left by the pin are measured using a micrometer fitted with a measurement rod.

Test probes are removed from the masonry by placing a small bearing sleeve over the probe and tightening a nut against the sleeve. Units can be damaged by spalls or splits during penetration tests and must be replaced in many cases.

5.2.4 *Analysis/Interpretation of Results*

Test results are analyzed by measuring penetration depths. The technique is most useful as part of a preliminary condition survey: areas exhibiting large probe penetrations indicate substandard or deteriorated materials. Thus results from penetration testing help determine areas requiring additional scrutiny using advanced investigative techniques.

Measured penetration depths may be used as an indication of material variation throughout the area being investigated but do not, by themselves, provide engineering information. Penetration depth can be calibrated against material properties using destructive tests, if desired.

5.2.5 *Summary of Probe Penetration*

Penetration tests can be conducted rapidly as part of a preliminary condition survey. Test results are used to provide an indication of the variation in material condition and deterioration. The test is not totally nondestructive and leaves small holes or spalls in tested units. Companion destructive tests on masonry units or prisms can be used to correlate penetration depth with material properties.

5.3 *PULLOUT*

Existing and installed masonry anchors are tested to determine tension and shear capacity by applying loads with a hydraulic jack. Anchor proof tests are required by many local building codes to verify design capacity. In addition, pullout tests of wedge-type, screw, or epoxy anchors carried out to failure have been used to provide a measure of masonry unit and mortar properties and to indicate masonry condition. Pullout strength can be correlated to masonry unit tensile strength through destructive laboratory tests of units or prisms. Two different procedures will be described: one for determination of proof loads for installed anchors and ties; a second method is used for indicating masonry unit properties using pullout strength of epoxied anchors.

5.3.1 *Background*

5.3.1.1 *Anchor Bolt Load Test*

The pullout strength of embedded or through-wall anchors is measured by applying a force to the end of the anchor. The force may be applied in direct tension, shear perpendicular to the axis of the bolt, or combined tension and shear, depending on the proposed function of the anchor. The maximum resisting force is usually used as a proof load to confirm bolt capacity.

The test can be carried out in a nondestructive manner if a displacement criteria is adopted for anchor bolt performance. The force required to cause a specified displacement of the free end is used as an indicator of anchor ultimate capacity. These tests must be carefully controlled to prevent failure of the masonry around the anchor.

Seismic strengthening provisions contained within the *Uniform Code for Building Conservation* (UCBC) [10] have been adopted by many municipalities. The UCBC stipulates that Standard 24-8 of the *Uniform Building Code* (UBC) [11] be followed when testing existing or installed wall anchors and bolts. ASTM Standards E 488-90, *Test Methods for Strength of Anchors in Concrete and Masonry Elements,* and E 754-80, *Test Method for Pullout Resistance of Ties and Anchors Embedded in Masonry Mortar Joints,* provide additional methods for pullout tests of installed masonry anchors.

5.3.1.2 *Masonry Pullout Strength*

Pullout strength also may be used as an indicator of material property variations. Anchors will usually fail the masonry in a combined ten-

Figure 5-5. Classical pullout failure of an embedded masonry anchor. The pullout force and the dimensions of the removed cone of masonry provide an indication of masonry strength.

sion/shear mode, pulling out a frustum or truncated cone of masonry around the anchor (Figure 5-5 a and b). Pullout strength of embedded anchors is used extensively in concrete construction to monitor strength gains of freshly placed concrete. Application of pullout techniques to evaluation of masonry is somewhat more difficult and, although the method may be appropriate, has not experienced widespread usage.

A complex stress field exists around anchors in tension, and the measured pullout strength cannot be used as a direct measure of compressive, tensile, or shear strength of masonry. Empirical relationships developed using companion destructive tests are used to provide an indication of material properties. Separate tests conducted on mortar and units can be analyzed to determine the strength of the masonry assemblage.

Application of pullout techniques to masonry evaluation has been the subject of several laboratory investigations. Victor Wilburn Associates used a modified version of concrete pullout equipment for masonry but concluded that the method is not especially useful for masonry evaluations. Pullout of threaded anchors epoxied into masonry units was correlated with masonry compressive and tensile strength by Noland [5]. The technique provided a fair degree of correlation with unit compressive strength but was marginal for use as an indicator of masonry tensile strength.

A variation of the pullout test using an expanding wedge-type sleeve [12] shows some promise for determination of mortar condition. An expanding sleeve is placed in a hole drilled into mortar and tightened against a small reaction ring, forcing a conical failure of the masonry around the sleeve. Strength can reportedly be estimated within 25% with no prior knowledge; calibration by tests on laboratory samples will provide a more accurate estimate of material properties.

The Building Research Establishment (BRE) in the U.K. has investigated the use of wedge-type anchors for pullout tests on clay and concrete masonry units, with only limited success [13]. The pullout strength of the masonry is measured by tightening a nut on the bolt against a reaction frame using a torque wrench. Initial trials show a reasonable correlation between the failure torque of the bolt and unit compressive strength, but a large variation in results prompted the BRE to investigate a more reliable pullout technique.

Trial pullout tests using a screw-type helix inserted into mortar provided more consistent results [19]. The test is simple and involves inserting a stainless steel screw into a mortar joint. The screw is driven into weaker mortars; a pilot hole is drilled for placement in stronger mortars and units. A supporting sleeve prevents buckling of the helix while allowing rotation as it is driven with a hammer. As the screw is pulled out a cylindrical annulus of material surrounding the helix fails in shear. The pullout force is measured by an apparatus that screws onto the projecting end of the helix. Results show a consistent relationship between pullout strength and mortar or unit compressive strength. With additional investigations this technique may become standardized for evaluation of masonry mortars and units.

Benli [14] conducted pullout tests where an entire brick was pulled from the surrounding masonry. The corresponding pullout force provided a measure of the mortar compressive strength with a correlation of 0.87.

Pullout tests are simple in principle and can be used as indicators of masonry material properties. The test can be conducted using basic hydraulic equipment and does not require extensive training of operating personnel. Applications of the method for masonry evaluation have been limited, however. The technique is not fully proven for masonry evaluation but has shown initial promise as an evaluation technique.

Pullout tests are somewhat destructive and have a tendency to leave cone-shaped pockmarks, spalls, and cracks that are not easily repaired. Tested units normally require replacement if appearance is of importance.

Pullout tests provide a local measurement of unit properties but do not indicate the interior condition of the masonry. Because of the localized nature of the test and large variability in test results a large number of tests are needed to provide a reliable estimate of overall masonry condition. The pullout test also is intended only for solid-unit or grouted masonry. The test would have to be modified for use on hollow or cored units to provide consistent results.

ASTM C 900-87, *Standard Test Method for Pullout Strength of Hardened Concrete,* describes pullout test procedures for use with concrete. The method uses metal inserts embedded in fresh concrete. This practice is obviously not possible with masonry; however, the loading equipment and general load procedure can be considered to be applicable to pullout tests of masonry.

5.3.2 *Equipment*

Pullout testing requires a hydraulic loading ram and pump, reaction frames, and devices for measurement of applied loads and displacements. Wedge-type, epoxied, and grouted anchors have all been used to determine masonry pullout strength. Epoxy anchors appear to provide the most consistent results; 1/4-inch-diameter threaded rods provide good bond to the epoxy, reducing the tendency for shear failure in the epoxy and are easily attached to the loading apparatus. High-strength structural epoxy is used to anchor threaded rods.

A hollow-core center-pull hydraulic jack is used for application of tensile loads. Use a load cell to measure applied forces. Alternatively, the hydraulic cylinder may be calibrated to determine load output as a function of internal pressure as measured with an attached pressure gage.

A small reaction frame is needed to distribute the applied load to the surrounding masonry. Reactions from the load jack must not interfere with the failure of the anchor: locate load reactions a distance equal to the embedded depth away from the anchor. Pullout strength will be affected by even slight misalignments between the axis of the jack and the bolt axis, hence a centering device and spherical seat as shown for the apparatus in Figure 5-6 are necessary.

Proof loads of installed anchors are based on the load causing a prespecified deformation at the loaded end. A dial gage or electronic device is used for measurement of these deformations. Tensile and/or shear loads are applied with hydraulic rams, reacting against a small steel framework attached to the masonry. Tensile loads also may be applied by turning a nut on the exposed bolt end using a calibrated torque wrench. For through-wall anchors, support frame reactions must be carried a distance equal to the wall thickness plus one-half the anchor plate width from the tested anchor.

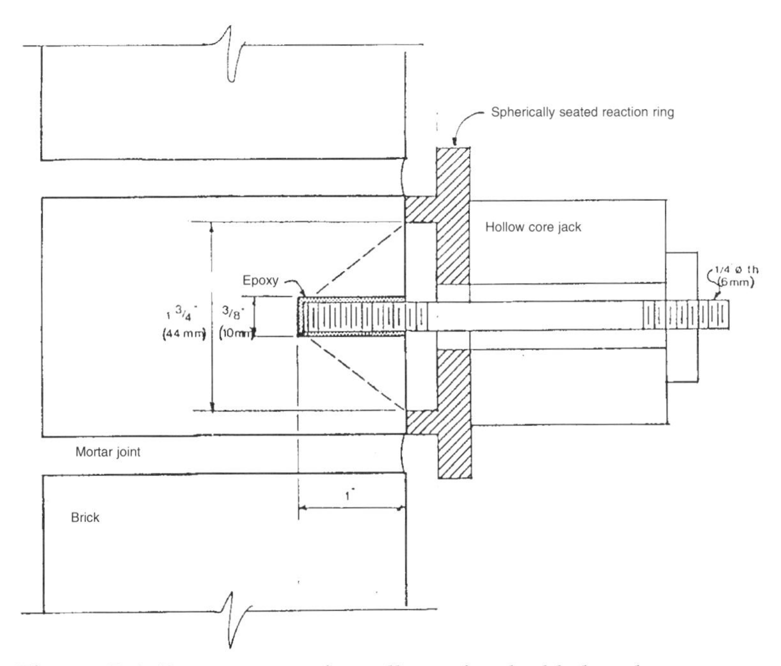

Figure 5-6. Test apparatus for pullout of embedded anchors.

Currently, no one manufactures pullout test equipment for masonry. The small reaction frames and necessary hydraulic equipment must be custom-assembled. A pulloff device for measurement of bond strength of concrete overlays and surface treatments is sold commercially, but its application to masonry testing is unknown. This device may be useful for measuring bond between face shells of hollow masonry and grout.

5.3.3 *Test Procedure*

Two different procedures are described here: one for determination of proof loads for installed anchors and ties; a second method is used for indicating masonry unit properties using pullout strength of epoxied anchors.

5.3.3.1 *Bolt Proof Load*

The UCBC requires a minimum of 10% of existing anchors be tested in pullout, with a minimum of four tests per floor. At least two tests are required on walls with floor joists framing into the wall and at least two additional tests must be conducted on walls with joists running parallel to the wall. The UCBC also requires that 25% of all new anchor bolts be tested, unless special inspection is provided during installation. ASTM E 488 specifies a minimum of three anchors of each size and type be tested. Determination of statistical data for calculation of mean bolt capacity requires testing between five and 30 anchors, depending on the coefficient of variation of test results (see Table 3, ASTM E 488). ASTM E 754 requires testing at least five masonry tie or anchor specimens for each combination of unit and mortar type.

Several procedures exist for testing to determine the tensile and shear strength of existing or installed masonry anchors. The UBC method for testing existing through-wall anchors [11] involves applying a tensile force to the bolt with a hydraulic jack. Reactions from the test apparatus are supported a distance equal to the wall thickness from the tested anchor. A pre-load of 300 pounds is applied to seat the anchor. Deformations are monitored during load application; the load that causes 0.125 inch relative movement of the bolt end is recorded as the proof load. The test is considered nondestructive because the anchor is not loaded to failure. The UBC provides a simple method for testing of newly installed anchors and bolts in combined tension and shear: a nut, installed over a malleable iron washer, is turned with a calibrated torque wrench. The UBC lists the minimum required applied torque for several different bolt diameters.

The procedure of ASTM E 488 is as follows: apply a small pre-load up to 5% of the estimated capacity to bring all members into full bearing. Apply loads in either an incremental or continuously increasing manner, taking displacement readings at regular intervals. The time of test is between two and 10 minutes. Failure is defined by one of several modes, including masonry failure in a shear-cone mode, pullout of the anchor, cracking of the masonry resulting in anchor pullout, and yielding or fracture of any portion of the anchor device. In addition, displacement criteria may be adopted to define anchor failure. The type of failure should be recorded along with the failure load.

ASTM E 754 describes two similar procedure for ties and anchors embedded in masonry mortar joints. Procedure A involves loading in a universal testing machine and is not readily applicable to field testing. Procedure B uses a center-hole load jack and cylindrical reaction ring, centered on the tie or anchor, similar to the apparatus shown in Figure 5-6. This procedure may be modified for field applications. Loading proceeds as described by ASTM E 488 (described above) until failure is observed. Failure in this case is one of three modes: pullout of the fastener due to bond failure between the tie or anchor and the mortar; failure of the mortar-unit bond adjacent to the fastener; or yielding or fracture of the fastener.

5.3.3.2 *Masonry Pullout Strength*

A technique using pullout tests of epoxy anchors as an indicator of masonry strength has been applied in the laboratory [5] and it may be possible to adapt the technique for field evaluations. Such tests for masonry pullout strength normally display a wide variation in results.

The research technique used a 5/16-inch-diameter hole drilled to a depth of 1 inch centered on the masonry unit to be tested. The inside surface of the hole was roughened using a standard 3/8-inch-diameter hardened steel tap to increase epoxy bonding to the brick. The hole was blown out using compressed air before insertion of a 1/4-inch-diameter threaded rod. The prepared hole was filled with epoxy and the threaded rod coated with epoxy. The rod was inserted to the full depth of the hole, with sufficient length projecting from the surface to allow attachment of the hydraulic jack.

Following sufficient cure time for the epoxy, the jack and hydraulic pressurization equipment was attached to the threaded rod. Deformation measurements were not taken during these tests. Failure occurred when the rod was pulled from the masonry, along with a small cone-shaped section of the unit (Figure 5-5). Different types of failure are possible as shown in Figure 5-7: specimens may fail by epoxy bond failure, fracture of the unit, or epoxy shear failure in addition to the desired pullout of a cone-shaped piece.

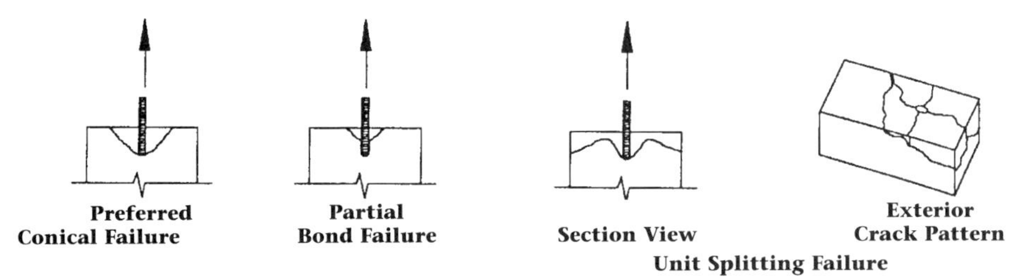

Figure 5-7. Several failure modes are possible when conducting pullout tests. Only the cone-shaped pullout failure can be correlated reliably to masonry material properties.

5.3.4 *Analysis/Interpretation of Results*

5.3.4.1 *Ties and Anchors*

The UCBC calculates allowable resistance values for existing anchors as 40% of the average of tension tests conducted as described above. According to the UCBC Commentary [15], this approach is based on failure tests of existing anchors and provides a safety margin of 5 against bolt failure. ASTM E 488 bases anchorage capacity on the average of three tests, providing no test result varies by more than 15% from the average. If any test exceeds the 15% allowable variation, two more tests must be conducted and the average of all five tests used. In this situation, the anchor capacity may also be based on the lowest of the original three tests. ASTM E 754 utilizes the mean of all tested strengths as the anchor capacity. Anchor capacities obtained by ASTM test methods must be further reduced by an appropriate factor of safety, depending on the requirements of local building ordinances.

5.3.4.2 *Masonry Pullout Strength*

Masonry pullout strength can be correlated to destructive tests on masonry units or mortar. Correlation tests conducted during laboratory

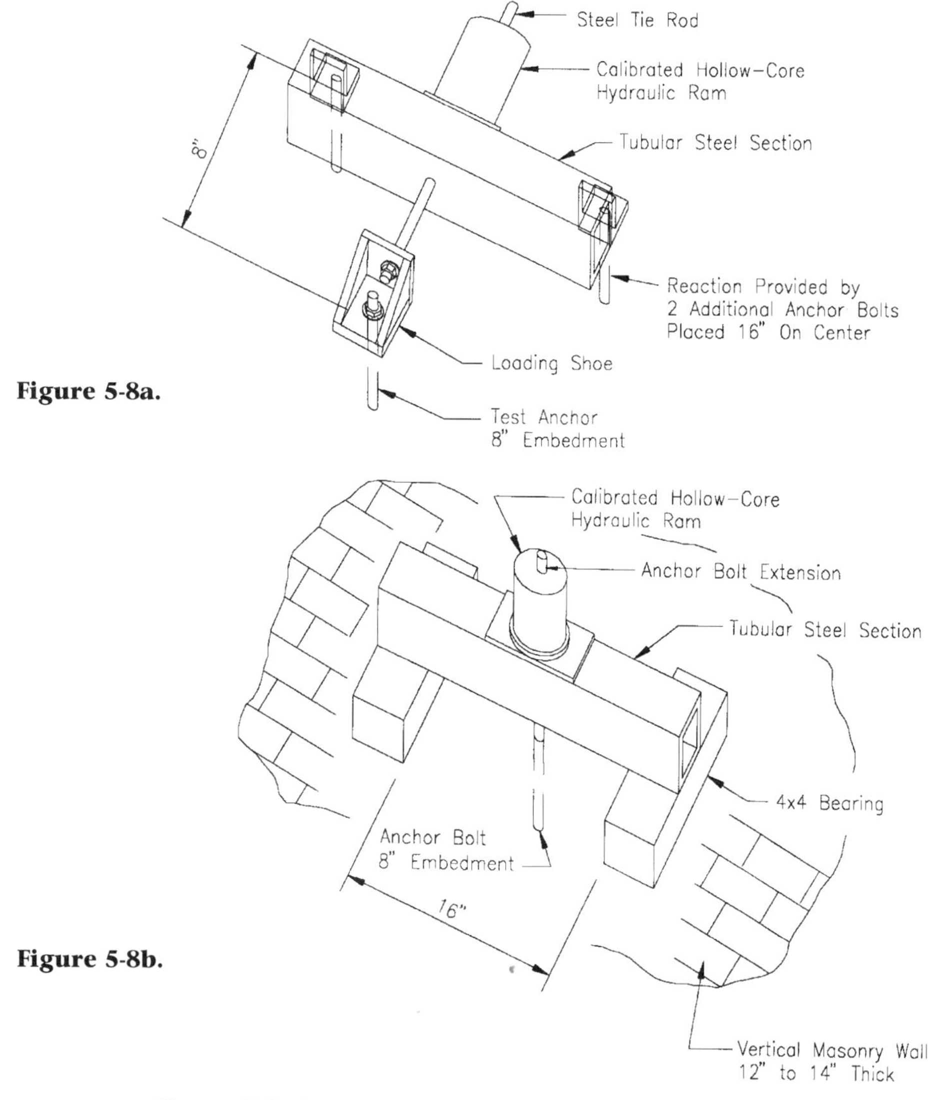

Figure 5-8. Apparatus used for (a) anchor bolt tensile pullout strength and (b) anchor bolt shear strength.

research [5] has found a correlation coefficient of 0.80 between pullout strength and unit compressive strength and a correlation coefficient of 0.6 between pullout strength and masonry unit tensile strength. Individual correlation studies must be conducted for each structure under consideration to develop empirical relationships between measured pullout strength and masonry material properties.

5.3.5 *Case Study*

5.3.5.1 *Anchor Bolt Tensile and Shear Resistance*

An historical building in New Orleans, Louisiana was rehabilitated for use as an office complex. A comprehensive condition assessment and structural evaluation determined the need for a supplemental steel structure within the masonry envelope to carry wind and gravity loads. The steel structure was to be attached to the masonry using epoxy anchors. The masonry itself was old and suffering from various stages of deterioration, and the anchor bolt manufacturer could not provide reliable information on bolt pullout and shear strength. A number of tests were conducted on bolts in the masonry to determine anchor tensility and shear resistance; the information was needed for the design of the anchoring system.

Installed bolts were secured with an epoxy-screen system and tested using one of two devices fabricated as a field adaptation of ASTM Standard Test Method E 488-90, *Test Methods for Strength of Anchors in Concrete and Masonry Elements*. The apparatus shown in Figure 5-8 was used to conduct a series of tension and shear tests on embedded anchors. A dial gage was used to measure anchor bolt displacements during load application and the tests were halted when:

- Failure from excessive cracking was observed
- Excessive deformations were recorded
- A pre-determined target load level equal to four times the maximum allowable bolt shear or tension load (according to the UBC for newly installed anchors) was surpassed

In the case of shear tests, bolts usually failed by crushing the brick in front of the anchor or by shearing the epoxy-brick interface during tension tests. In some situations, surrounding mortar failed resulting in the movement of the entire brick.

5.3.6 *Summary of Pullout*

Pullout techniques are useful for determining proof loads of existing and installed masonry anchors. Test results are used in the determination of anchor design capacity values. Pullout tests may be nondestructive if a displacement criteria is adopted for defining anchor performance, and loading is terminated before fracturing the bolt or surrounding masonry.

Pullout of embedded anchors has been correlated to masonry tensile properties but is not normally recommended because of the destructive nature of test and questionable usage of results. The technique offers an alternative method for determination of material uniformity; however, it must be used with caution until additional supportive data is assembled.

5.4 REFERENCES

1. Noland, J.L., R.H. Atkinson, and J.C. Baur, Atkinson-Noland & Associates Inc. 1982. An Investigation into Methods of Nondestructive Evaluation of Masonry Structures, National Technical Information Service Report No. PB 82218074. Report presented to the National Science Foundation.
2. International Society for Rock Mechanics. 1978. Suggested Methods for Determining Hardness and Abrasiveness of Rocks. *International Journal of Rock Mechanics, Mineral Science and Geomechanics*. Vol. 15.
3. Natrella, M.G. 1963. *Experimental Statistics*. National Bureau of Standards Handbook 91.
4. Donovan, J.E. 1991. Flexural Tensile Bond Strength of Masonry Prisms. Master's thesis, University of Wyoming, Laramie.
5. Noland, J.L., R.H. Atkinson, G.R. Kingsley, and M.P. Schuller, Atkinson-Noland & Associates Inc. 1990. Nondestructive Evaluation of Masonry Structures. Reported presented to the National Science Foundation.
6. ASTM. 1975. *Standard Test Method for Rebound Number of Hardened Concrete,* ASTM C 805-75. Philadelphia: ASTM.
7. Victor Wilburn Associates. 1979. Nondestructive Test Procedures for Brick, Block and Mortar. Report to the Department of Housing and Urban Development, Contract No. H2540.
8. RILEM. 1993. *MS.D.7, Determination of Mortar Surface Hardness by Pendulum Hammer.* Paris: The International Union of Testing and Research Laboratories for Materials and Structures.
9. ASTM. 1993. *Standard Test Method for Penetration Resistance of Hardened Concrete,* ASTM 803-90. Philadelphia: ASTM.
10. International Conference of Building Officials. 1991. *Uniform Code for Building Conservation.* Whittier, Calif.: International Conference of Building Officials.
11. International Conference of Building Officials. 1991. *Uniform Building Code Standard, No. 24-8, Tests of Anchors in Unreinforced Masonry Walls.* Whittier, Calif.: International Conference of Building Officials.
12. Domone, P.L., and P.F. Castro. 1987. An Expanding Sleeve Test for In Situ Concrete and Mortar Strength Evaluation. Proceedings at the International Conference on Structural Faults and Repair, London.
13. de Vekey, R.C. 1991. In Situ Tests for Masonry. Proceedings at the 9th International Brick/Block Masonry Conference, Berlin, Germany.
14. Benli, G., F. Wei, and W. Ning. 1991. The In Situ Test Methods of Bonding Mortar. Proceedings at the 9th International Brick/Block Masonry Conference, Berlin, Germany.
15. Structural Engineers Association of California. 1992. Commentary, Appendix Chapter 1. In the *Uniform Code for Building Conservation, Seismic Strengthening Provisions for Unreinforced Masonry Bearing Wall Buildings*. Los Angeles. SEAOC.
16. ASTM. 1993. *Standard Test Method for Pullout Strength of Hardened Concrete,* ASTM C 900-87. Philadelphia: ASTM.
17. ASTM. 1993. *Test Methods for Strength of Anchors in Concrete and Masonry Elements,* ASTM C 900-87, E 488-90. Philadelphia: ASTM.
18. ASTM. 1993. *Test Method for Pullout Resistance of Ties and Anchors Embedded in Masonry Mortar Joints,* ASTM C 900-87, E 754-80. Philadelphia: ASTM.

19. Ferguson, W.A., and J. Skandamoorthy. 1994. The Screw Pull-Out Test for the In Situ Measurement of the Strength of Masonry Materials. Proceedings at the 10th International Brick and Block Masonry Conference, Calgary, Alberta, Canada.

Chapter Six • *Shear Testing*

Masonry shear strength is an important material property when determining resistance to seismic and wind loads. Recent interest in retrofit and seismic upgrade of older unreinforced masonry buildings has increased the need for reliable information on shear resistance. Characterization of masonry shear strength is essential when lateral load resistance is of importance, and in some seismic areas the measurement of shear strength is required for retrofit designs. Masonry shear strength does not correlate well with in-place measurements such as pulse velocity [1] and hence direct measurement of shear strength using destructive or temporarily destructive procedures is necessary.

Shear capacity is difficult to measure in-place and most currently available techniques do not provide an absolute value for design purposes. In-place methods are useful, however, for providing a comparative measure of masonry shear strength and can be correlated to full-scale wall behavior using empirically derived relationships.

Several tests can be conducted for the determination of masonry shear strength; however, most of these techniques are quite damaging and require significant repairs after testing. Large-diameter cores may be removed and tested in compression machines to determine bed joint shear strength as shown in Figure 6-1 (a). The cylindrical space remaining after removal of cores is difficult to repair. Wall sections can be removed from the wall for laboratory testing (Figure 6-1 (b)) — a costly and difficult process.

In-place shear tests can provide reliable information on masonry bed joint shear capacity. Large-scale tests may load entire piers or portions of shear walls in-place as shown in Figure 6-1 (d) to obtain masonry shear strength and deformation properties. Determination of shear strength using the in-place shear, or push, test (Figure 6-1 (c)) minimizes disturbance to the masonry being investigated and is the preferred

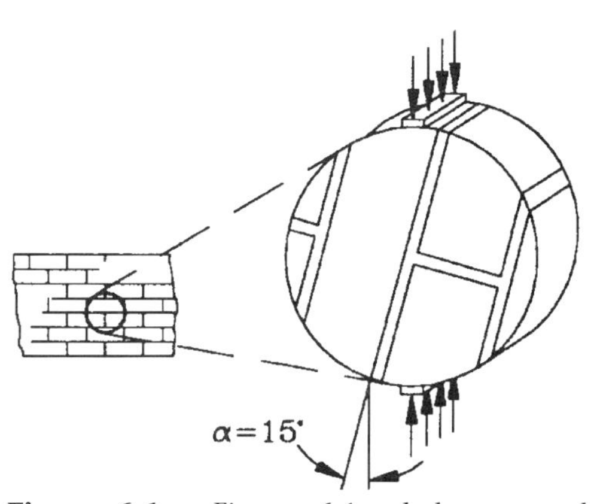

Figure 6-1a. Figures 6-1 a-d show several destructive and in-place tests are available for determining masonry shear strength, including an 8-inch diametral core shear test.

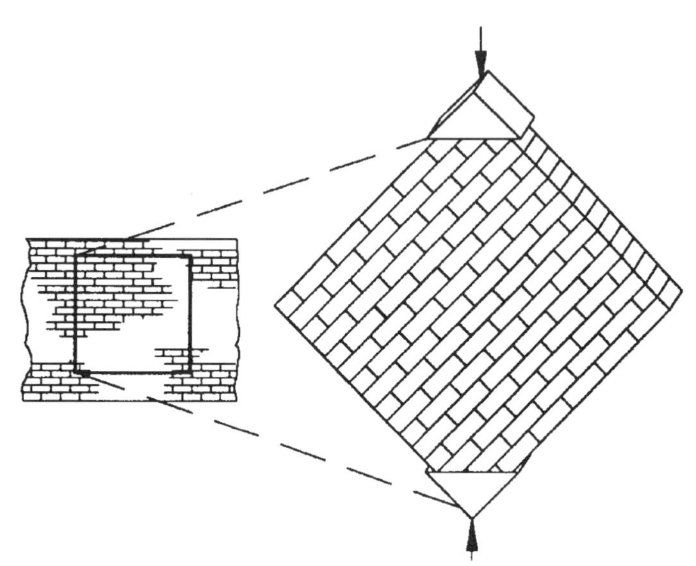

Figure 6-1b. The ASTM E 519 diagonal tension (shear) test.

technique for shear testing of unreinforced masonry buildings. The test is simple to perform and involves measuring the load required to displace a single masonry unit. The in-place shear test provides a measure of masonry bed joint shear strength. This strength index is not intended to correlate with masonry compressive strength and can be used to predict overall masonry structural shear strength only through the use of empirically derived relationships.

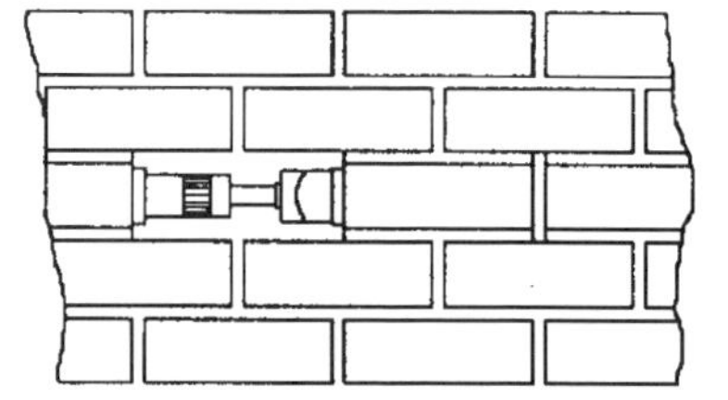

Figure 6-1c. The in-place shear (push) test.

6.1 *BACKGROUND*

Flexural cracking and compressive crushing may occur in masonry piers and shear walls subjected to in-plane loads; however, masonry shear strength typically controls overall resistance. Increasing lateral loads precipitate shear failure by sudden formation of diagonal cracks as shown in Figure 6-2. In old masonry characterized by stiff units set in weak lime-based mortar, diagonal cracks usually propagate through the mortar joints with the crack taking on the commonly observed stairstep pattern (Figure 6-2 (a)). In modern masonry with stronger cement-based

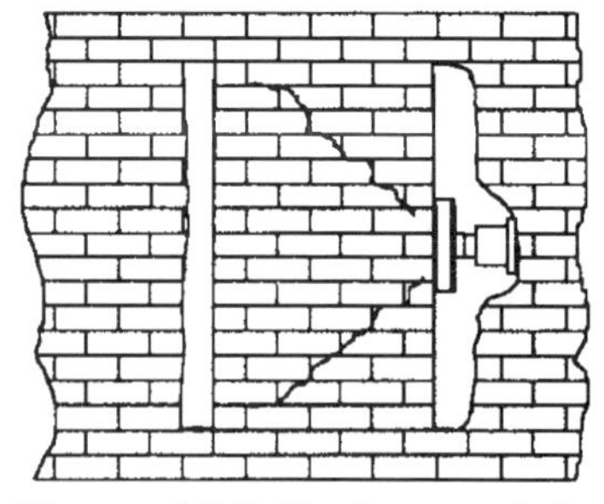

Figure 6-1d. The large-scale in-place shear test.

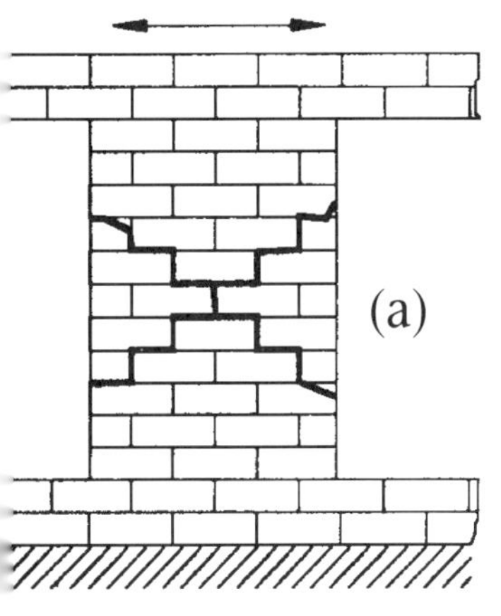

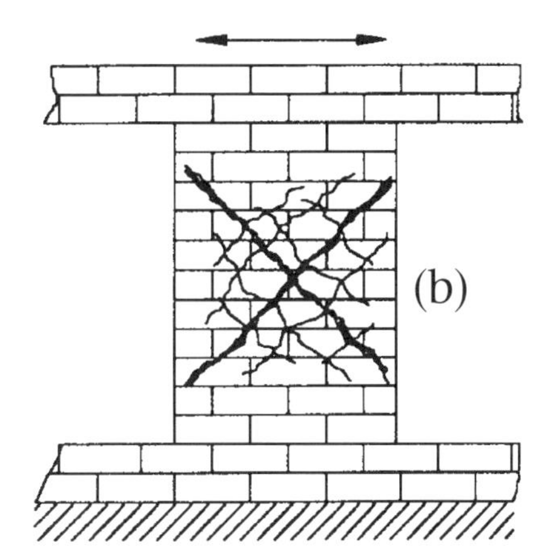

Figure 6-2. Shear crack development in (a) unreinforced masonry and (b) reinforced masonry.

mortars and grouted reinforced masonry, the cracks will propagate fully through the units in a diagonal direction (Figure 6-2 (b)).

Masonry shear resistance is a complicated subject and is not fully understood even though shear behavior normally governs the response during lateral loadings. A complex state of stress exists before formation of a shear crack consisting of a combination of non-uniform compression or tension stresses normal to the developing crack and shear stresses along the crack. Hence shear behavior is difficult to measure without resorting to large-scale testing. Small-scale tests are applicable only because they have been correlated to full-scale masonry response and most provide little information by themselves.

Shear deformability also can be difficult to measure. Many advanced analytical techniques such as the finite element method require accurate information regarding shear modulus G; however, this value can be difficult to measure using standardized tests. A method for determination of the shear modulus based on deformations of masonry piers tested in-place is described later in this chapter.

6.2 *IN-PLACE PUSH TEST*

The in-place shear test, or push test, is the most widely accepted method for determining masonry shear strength. The test uses an empirical relationship between in-plane shear capacity of a masonry structural element and in-place shear capacity measured by displacing horizontally a single brick. This value is modified by several factors to account for assumptions regarding workmanship variability, the contribution of mortar in the collar joint to the measured shear load, and the effect of stresses normal to the tested joints. The method is described in several masonry building codes and is required for determining seismic resistance of existing unreinforced masonry buildings in many municipalities throughout the state of California.

The in-place push test is simple to conduct, requiring removal of a single masonry unit and a vertical head joint on opposite sides of the chosen test unit. The basic test setup is shown in Figure 6-3. The test unit is displaced horizontally using a hydraulic cylinder or small flatjack, thus providing an indication of mortar quality, and internal cohesion of the mortar.

The test does not provide an absolute design value for masonry shear strength and must not be applied in this manner. Results from the push test are related to overall structural masonry shear strength using empirical relationships to provide a basis for design shear resistance.

The in-place push test can be correlated to overall masonry shear strength because of the manner in which masonry shear failure propagates. Most masonry piers and shear walls have a small aspect ratio, with wall length being equal or greater than the height. These short, squat masonry panels will typically fail by formation of a diagonal shear crack. Older masonry structures with relatively weak mortar usually fail through the mortar head and bed joints in the familiar stairstep pattern (see Figure 6-2 (a)). Tensile capacity of vertical head joints is normally very small, hence the main resistance to shear failure is provided by horizontal bed joints. Failure propagates by fracture and horizontal sliding along bed joints, thus providing the relationship between the sliding resistance of a single masonry unit and overall wall shear strength. Shear strength of tall, narrow piers and columns, which fail in flexure by rocking actions, cannot be predicted using the push test.

The in-place push test is used, therefore, to determine an experimental value related to mortar quality and internal mortar cohesion. This value is related to the failure shear force of the masonry using an empirically derived formula. This relationship was determined during development of the ABK Methodology [2] by relating push test results with tests conducted on full-scale wall elements subjected to lateral loads. The recommendations of the ABK Methodology were adopted by the International Conference of Building Officials and forms the basis for determination of masonry shear strength described in Appendix Chapter 1 of the Uniform Code for Building Conservation [3].

Modern masonry with strong, cement-based mortars and grouted reinforced masonry subjected to lateral loadings normally does not fail by stairstep-type cracking described above. Shear cracks instead propagate in a diagonal direction directly through individual units with little regard for the location of mortar joints (Figure 6-2 (b)). Hence the push test does not provide a reliable estimate of shear capacity for this failure mode and should not be used for this type of masonry. In such cases it is recommended that other techniques such as the diametral core test or large-scale in-place tests for masonry shear strength (described below) be used.

6.2.1 *Equipment*

The in-place push test is simple to conduct and requires only the most basic equipment. The test setup is shown in Figure 6-3. Necessary equipment includes the following:

- Rotary drill and masonry bits for mortar removal
- A variety of hand chisels and hammer

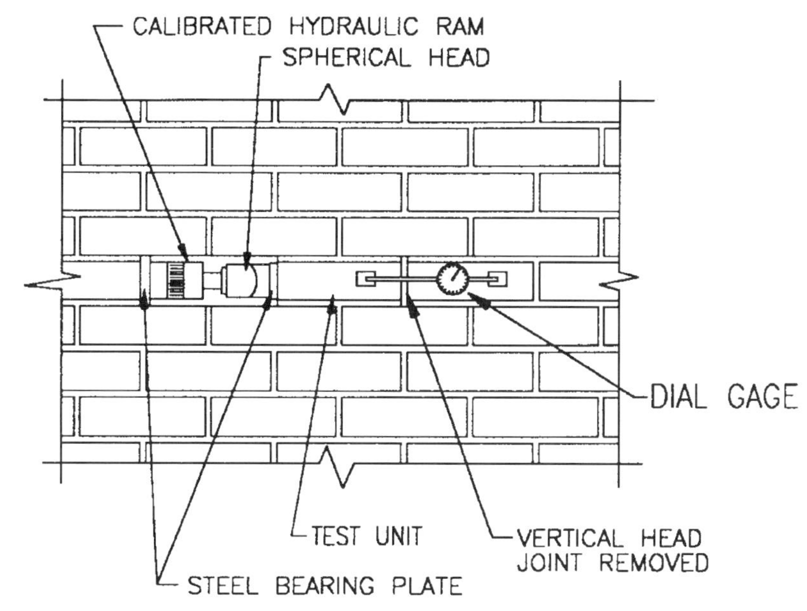

Figure 6-3a. Test setup for the in-place shear test using a hydraulic cylinder for application of shear forces.

- Calibrated hydraulic ram or flatjack
- Pressure gage
- Hand pump
- 1/2″-thick steel loading plates
- 2″-diameter spherically seated bearing block
- Metal shims or spacers
- Eye and face protection

An electronic pressure transducer or load cell may be substituted for the pressure gage for measuring applied loads. It also is desirable to provide a positive means for measurement of unit displacements. Dial gages or electronic displacement transducers may be mounted to the test unit for this purpose. Use of electronic equipment require attachment to a data acquisition device and portable computer.

6.2.2 *Procedure*

Descriptions of the in-place push test are described in *The Uniform Code for Building Conservation* (UCBC) [3] and *The Uniform Building Code* (UBC Standard No. 24-7, In-Place Masonry Shear Tests) [4]. Committee C-15 of the ASTM is currently developing a standard test method for the push test. A step-by step description of the method is included here.

The number of tests required by building codes varies from source to source. Most requirements are similar to those stipulated in the UCBC, which requires a minimum of one test per 1,500 square feet of wall sur-

Figure 6-3b. Test setup for the in-place shear test by insertion of a small flatjack into a head joint for application of shear forces.

face and at least one test per wall or line of wall elements resisting lateral forces. Two tests per wall or line of wall elements are required at each of both the first and top stories. In any case the total number of tests per structure must be greater than eight.

Test locations should be chosen to represent any variations in material quality, workmanship, weathering, and deterioration that may be present. Areas in which bed joints are not parallel should be avoided as this may lead to erroneously high resistance resulting from a wedging action of the test brick. Cracked or broken units also should be avoided. Place the loading jack to bear against a sufficient mass of masonry to prevent unanticipated failure of the masonry behind the jack. Do not test in areas directly adjacent to wall ends or openings.

At the chosen test location it is necessary to isolate the test joints. A single unit on one side of the test unit is removed for placement of the hydraulic loading cylinder. The head joint on the opposite side of the unit also must be cleared. Mortar can be easily removed by successively overlapping holes (stitch drilling) using a masonry bit with a diameter slightly less than the thickness of the mortar joint. The removed unit should be set aside for replacement in the wall after completion of the test.

All head joint mortar must be removed from both sides of the test unit, as shown in Figure 6-3 (a), to permit free movement of the upper and lower surfaces of both the top and bottom bed joints. Dimensions of the top and bottom bed joints are measured and recorded. It is recommended that the percent coverage of collar joint mortar (behind the test unit) be estimated and recorded, as this area of mortar also contributes to initial load resistance.

A hydraulic cylinder is inserted into the space and shimmed to center the point of load application on the brick end. A 1/2-inch-thick steel bearing plate is placed between the jack and the test unit. Dimensions of the steel plate should be slightly less than the dimensions of the loaded face and should be placed such that no load is applied to either the top or bottom mortar joints. Placing a small spherically seated bearing block between the load jack and test unit permits more effective testing when jack bearing surfaces are slightly misaligned.

A small flatjack inserted into a cleared head joint also can be used to load the test unit. The flatjack has dimensions equal to or slightly less than the end dimensions of a typical masonry unit. The jack is inserted into the head joint as shown in Figure 6-3 (b) and is pressurized hydraulically. This approach is attractive because less labor is required. Loading with a hydraulic cylinder requires removing an entire test unit; a flatjack insertions requires clearing a single head joint. This also eases post-test repair work and provides a uniform loading of the test brick. The main disadvantage is that it is not possible to determine mortar coverage in the collar joint as required by the UCBC because an entire masonry unit is not removed. Flatjacks also have a limited deformation range, and it can be expensive to manufacture a different flatjack for each unit size. See Chapter 9 for more information on calibration and use of flatjacks for masonry testing.

The method requires that the "load at first movement" of the test unit be recorded as the failure load, as indicated by relative movement associated with bed joint shear failure. "First movement" of the test unit is often difficult to define and it is advantageous to affix dial gages or electronic displacement transducers to monitor displacement of the test unit. Locate displacement measuring devices as shown in Figure 6-3 (a).

Loads are applied gradually, at a rate such that joint failure occurs within one to three minutes after initiation of the loading sequence. The load at first movement of the test unit is recorded.

In many cases the load continues increasing as the test unit displaces (Figure 6-4). Dilation of the mortar within the joints following failure has the effect of slightly increasing the joint normal stress, thus artificially causing an increase in joint shear resistance. In such cases it is advisable to use the load at which there is a break in slope of the load-displacement curve as the so-called "first movement" load as shown in Figure 6-4.

If no displacement instrumentation is available, a finger placed across the joint being sheared can sense a movement of about 0.01 inch, indicating shearing of the mortar joint. ASTM E6 *Definitions of Terms relating to Methods of Mechanical Testing* describes several techniques for determining the yield point of materials. The deflection offset technique, using a deflection of about 0.01 inch, is appropriate for shear testing.

The magnitude of normal stress on the test joint has a direct effect on resistance to movements. Hence it is necessary to determine the state of vertical compressive stress at

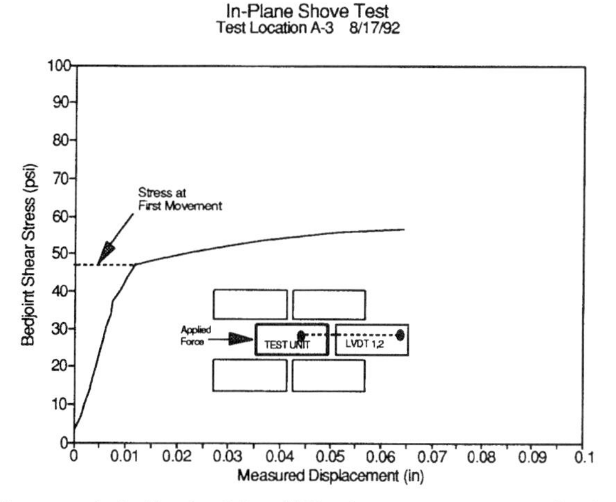

Figure 6-4. Typical load/displacement curve obtained during an in-place shear test.

the test location, either by estimation based on assumed dead and current live loads or by measurement using flatjack techniques described separately in Chapter 9. The magnitude of normal stress on the tested joints must be recorded for each location.

Following completion of the test all instrumentation and loading equipment is removed from the wall. The unit removed for placement of loading apparatus is replaced in the wall. Mortar is packed around this unit and in the cleared head joint to restore the wall to its original appearance. It is usually desirable to drill out and replace the failed mortar around the test unit.

6.2.3 *Analysis of Test Results*

The load recorded at a relative movement associated with bed joint failure is useful as an indicator of relative mortar quality. This value also may be used to estimate masonry shear strength using empirical relationships that relate the shear strength of a single unit to the shear strength of unreinforced masonry walls subjected to lateral loads. The tested bed joint shear strength is calculated as follows:

$$v_{test} = \frac{P}{A_j} \qquad (6\text{-}1)$$

where P is the measured applied load at first movement of the test unit and A_j is the total area of the top and bottom tested joints.

The general analysis procedure is described in the ABK *Methodology for Mitigation of Seismic Hazards in Existing Unreinforced Masonry Buildings* [2]. The methodology reduces the measured bed joint shear strength to the shear strength at zero normal stress, v_{t0}, using a Mohr-Coulomb type of relationship:

$$v_{to} = v_{test} - \phi \frac{P}{A} \qquad (6\text{-}2)$$

In this case the area A is the average area of the upper and lower tested bed joints. This relationship considers the measured shear strength to be equal to a mortar cohesion value that occurs at zero normal stress plus a frictional component, which depends on the magnitude of compressive stress normal to the tested bed joints. The friction coefficient ø is assumed to be 1.0 based on experimental observations. The term P/A represents vertical stresses due to any existing dead and live loads at the test location.

The shear strength v_{t0} is determined for each individual test. From these values a basic bed joint shear v_t is determined as the 20th percentile of tested values, that is, the test value at which 80% of tested values is greater (see Chapter 1 for a discussion of statistical methods). It is this value for tested bed joint shear strength v_t, which is used to determine allowable shear stresses for unreinforced masonry. The UCBC requires that mortar with a 20th percentile shear strength, v_t, less than 30 psi be repointed with new mortar and retested.

Different levels of allowable shear stress values are recommended by different design codes; however, all formulae follow the same basic approach. The allowable shear stress v_a is calculated as:

$$v_a = k\left(rv_t + \phi \frac{P}{A}\right) \tag{6-3}$$

The constant k is an empirically derived coefficient that adjusts measured bed joint shear to overall pier shear resistance accounting for workmanship variability. The reduction factor r adjusts the tested values for the contribution of the collar joint to the measured bed joint shear strength, and v_t is the 20th percentile basic bed joint shear strength measured by in-place testing. ø is a factor to account for increases in pier shear resistance resulting from the existing axial stress *P/A* which acts normal to the shear surface. Large-scale tests on masonry piers conducted during development of the ABK Methodology have been correlated with in-place push tests to determine appropriate values for *k, r,* and *ø* . The ABK methodology recommends k=0.75, r=0.75, and ø=1.0 which results in:

$$v_a = 0.75\left(0.75v_t + \frac{P}{A}\right) \tag{6-4}$$

The UCBC contains an expression for determination of allowable shear stress:

$$v_a = 0.1v_t + 0.15\frac{P_d}{A} \tag{6-5}$$

which is similar to the ABK formula but considers only the permanent dead load P_d during calculation of vertical stresses. The *UCBC* also stipulates that the allowable shear stress v_a can be no greater than 100 psi.

The in-place push test is based on basic engineering mechanics and has been proven to correlate well with masonry pier behavior for most cases. Subsequent research by Schmid [5] and Abrams [6] show that the push test can overestimate wall shear strength by up to 300% in some cases. Large variations in test values also can be expected, even for tests conducted in the same immediate area. A coefficient of variation of 20% to 30% or more can be expected for push test data. Some amount of error can be attributed to the highly variable nature of masonry construction; however, several of the simplifying assumptions inherent to the test serve to increase variability in test results. These assumptions can be addressed by modifications to the test procedure discussed in the following sections.

6.3 *IN-PLACE PUSH TEST CONDUCTED WITH FLATJACKS TO CONTROL VERTICAL STRESS*

The in-place push test has been related to overall masonry shear strength and provides the most rational basis for determination of design shear values currently available. There are, however, several basic assumptions

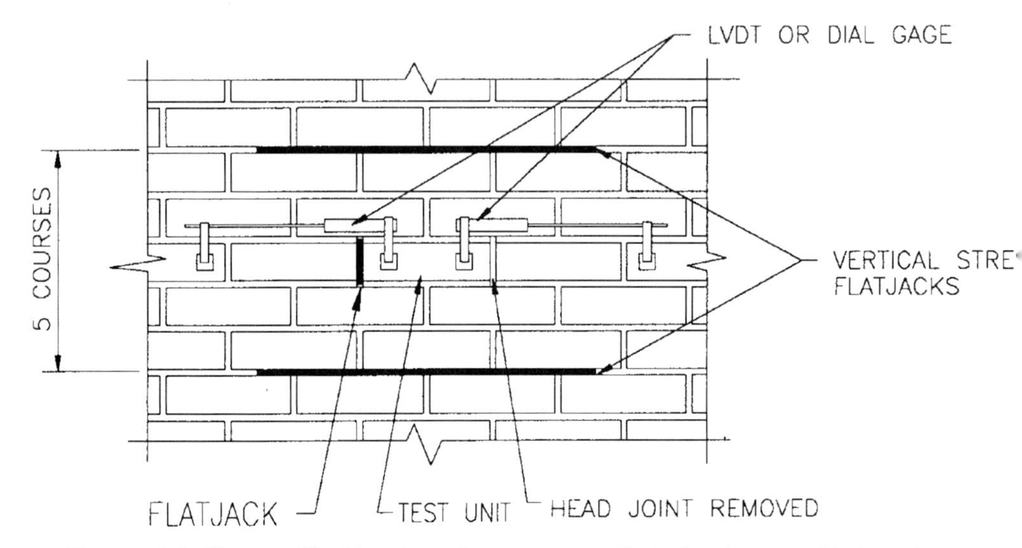

Figure 6-5. The modified in-place shear test uses flatjacks above and below the test brick to regulate the magnitude of stress normal to the tested bed joints.

that need to be addressed to fully understand the implications of relating bed joint shear test values to masonry shear strength. Results from in-place shear tests normally have a large variation and any procedural techniques that reduce these variations would be most advantageous.

Flatjack tests in conjunction with the in-place shear test can be used to control normal stress on the test unit (Figure 6-5), reducing uncertainties regarding existing vertical stresses and providing a method for determining the relationship between joint normal stress and joint shear strength. Use of flatjacks during the in-place push test requires little effort if the push test is conducted following an in-place deformability test at the desired test location. The flatjacks remain in place following the deformability test and can subsequently be used to control normal stresses during the push test. For this situation loads applied during the deformability test must be restricted to a level which does not cause damage to mortar joints and units involved with the push test.

Use of flatjacks also ensures the stress normal to the tested joints remains constant throughout the test. Mortar in masonry joints has a tendency to dilate, or increase in volume, resulting from even small shear deformations. If this dilation is restrained by the masonry surrounding the test joint, the net effect will be a slight increase in the joint normal stress resulting in a greater measured shear value. Maintaining flatjack pressure at a constant level during shear testing allows joint dilation to occur while keeping the joint normal stress at a constant level.

The standard UBC test method requires reduction of the tested bed joint shear strength to the shear strength occurring at zero compressive stress normal to the bed joints based on a Mohr-Coulomb criteria:

$$v_{to}=v_{test} - \phi \frac{P}{A} \quad (6\text{-}6)$$

This formula utilizes the friction coefficient ø, which is assumed to be equal to one for the ABK Methodology based on measurements of masonry tested in the Los Angeles area. This assumption is valid; however, it must be recognized that masonry in other parts of the country may vary and the friction coefficient can be considerably different from the assumed value of 1.0. Laboratory studies [7,8,9] have shown that the friction coefficient for masonry mortars can vary from 0.37 to 1.15, with a mean value of approximately 0.7. Hence it may be necessary to determine the friction coefficient for masonry that is significantly different than that tested in the ABK study.

By using flatjacks to control joint normal stress the coefficient of friction can be determined rapidly for several locations in the structure being investigated. The simpler push test can then be conducted throughout the remainder of the structure and data reduction performed using the experimentally determined friction coefficient in Equation 6-6.

6.3.1 *Equipment*

The modified in-place push test utilizes all equipment necessary for the basic push test described in the previous section: drill plus masonry bits for mortar removal, flatjack with appropriate dimensions to fit in a head joint for application of shear loads, and deformation measurement devices. Equipment requirements, installation, and usage of flatjacks follows the procedures outlined in the chapter on flatjack testing. The main difference is that two hydraulic pumps are required: one to maintain flatjack pressure and the other to control the horizontal loading flatjack. The two flatjacks for vertical stress application are connected in parallel such that they are pressurized equally at all times.

6.3.2 *Procedure*

The test equipment is installed in the configuration shown in Figure 6-5, with two parallel flatjacks separated by five courses of brick masonry. The unit for shear testing should be located on the middle course and centered between the flatjacks; it is preferable to use a flatjack inserted into a cleared head joint for application of shear loads. This approach requires removal of only two head joints, one on either side of the test unit, and ensures a fairly uniform distribution of vertical stresses across the test unit as applied by flatjacks above and below the test unit. The classical technique of removing an entire brick for insertion of loading devices can have a significant effect on vertical stresses, and can lead to stress concentrations and a non-uniform vertical stress distribution along the tested joints.

The initial loading cycle is conducted with zero pressure in the vertical stress flatjacks by increasing pressure in the head joint flatjack. The load at first movement of the test unit is recorded and provides the base shear strength v_{to} occurring with zero normal stress on the tested joints. Following this sequence, the pressure in the vertical stress flatjacks is increased and the test unit loaded until it again begins to move. The loading sequence is repeated for several increments of increasing normal stress. Figure 6-6 shows a load-displacement curve obtained by conducting a modified in-place shear test at several increments of normal stress. Several increments of 25- to 50-psi joint normal stress are necessary to accurately describe the joint failure surface for determination of the friction coefficient. If desired, the loading jack may be removed and placed in the void on the opposite side of the test unit to conduct a cyclic push test.

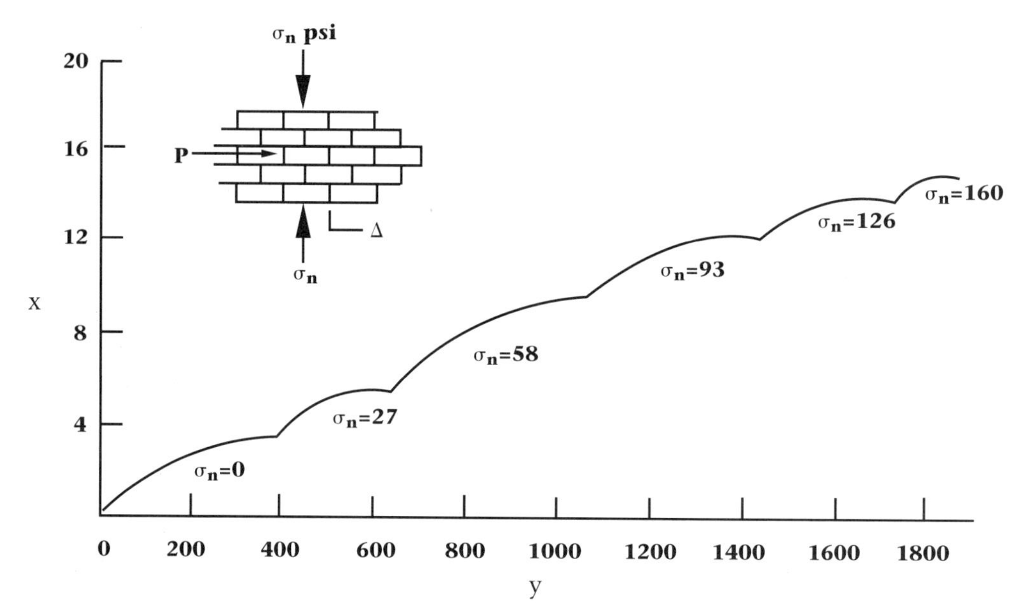

Figure 6-6. Load/displacement curves obtained during a modified in-place shear test for increasing levels of normal stress.

Following completion of all test sequences the equipment is removed from the test site. The removed masonry units are replaced and flatjack slots repointed using a mortar similar in color and composition to the original mortar.

6.3.3 *Analysis of Test Results*

The joint failure stress is calculated based on the total bedded shear area as described above and tabulated for each value of normal stress. The bed joint shear strength at zero normal stress is determined by the failure load at the initial loading sequence (where the vertical stress flatjack pressure was zero) and used for determination of design shear stress values as described in the previous section.

When a small flatjack inserted into a cleared head joint is used to

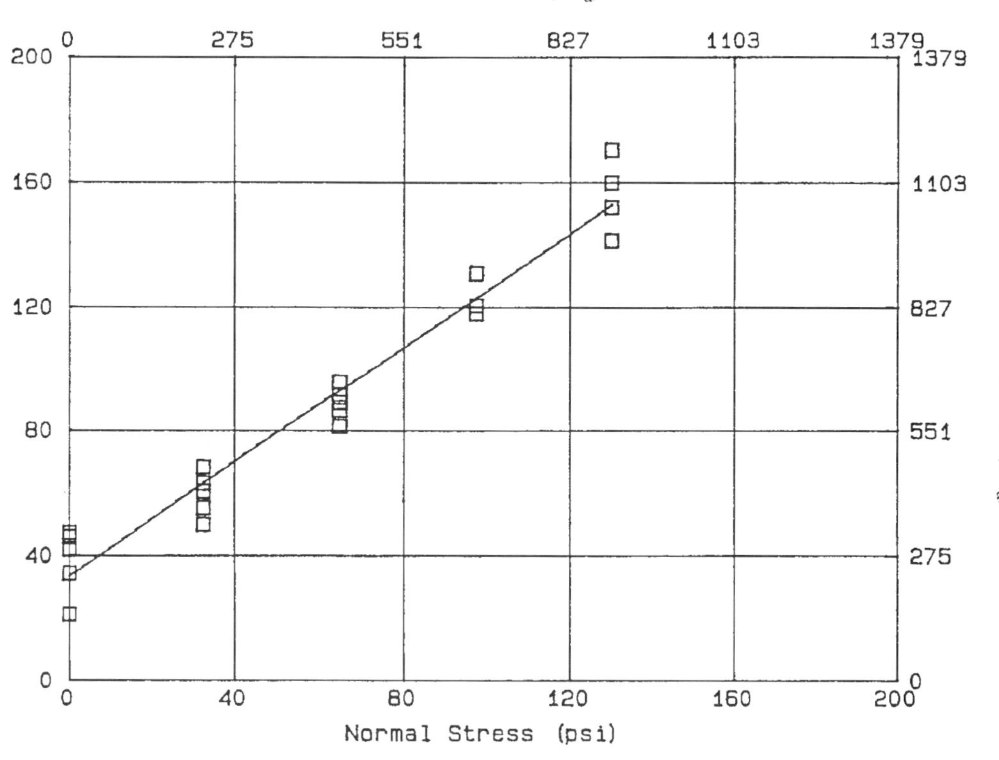

Figure 6-7. The Mohr-Coulomb failure surface for a mortar joint obtained during the modified in-place shear test. The slope of the line is the friction angle ϕ.

apply shear forces to the test unit, the magnitude of vertical stress on the test joints can be considered equal to the stresses applied to the masonry by the vertical stress flatjacks. See Chapter 9 for guidelines on reducing flatjack pressure to masonry stress.

The Mohr-Coulomb failure surface is determined by plotting the bed joint failure shear for each level of normal stress as shown in Figure 6-7. A line fit to the data points using linear regression provides the failure surface; the slope of this line is the friction coefficient μ.

In a normal investigation the friction coefficient μ is determined at several test locations and an average value is determined to be representative for each combination of mortar and unit type present in the structure. Analysis of data from subsequent tests conducted without using flatjacks may utilize this value for the friction coefficient in Equation (6-6).

Thus the modified in-place shear test, using flatjacks to control stresses normal to the tested joints, accounts for two conditions that may contribute to measured variations in mortar quality: the magnitude of normal stress and friction coefficient μ. Direct measurement of these values increases reliability of test results.

6.4 DIAMETRAL CORE TEST

A method for determination of shear strength using large-diameter cores removed from existing masonry has been used for several years [10]. This technique is most useful for use with modern masonry and structures with relatively high-strength mortars where the in-place shear test is not feasible or appropriate. Until recently the diametral core test was allowed as an alternate to the in-place push test by many building codes however the method is now obsolete for most applications with the push test being the preferred technique for determining masonry shear resistance. Core testing is also not applicable to all types of masonry. It is difficult to remove intact cores from older masonry with weak lime-based mortars; in these cases it is best to use the push test described above.

The test method involves loading a core specimen in compression at an angle of 15 degrees from a diametral masonry bed joint, as shown in Figure 6-8. The average shear strength of the joint τ can be calculated as:

$$\tau = \frac{P\cos\alpha}{A} \tag{6-7}$$

where P is the measured compressive load at failure, α is the inclination of the bed joint to the loaded axis (15°), and A is the area of the test joint.

The peak stress that initiates failure may be significantly different from the average shear stress calculated using Equation 6-7. Loading of a core specimen in compression produces a combination of normal and shear stresses at the test joint and results in a compressive stress normal to the sheared joint which is highly nonuniform. The failure load also is quite sensitive to minor variations in the angle α: an alignment error of plus or minus two degrees will result in a variation in the peak stress of 7% (for α=17 degrees) to 14% (for α=13 degrees) [19]. In addition, several different failure modes may occur: shear failure within the mortar, shear along the brick-mortar interface, tension failure in the brick or mortar, or any combination of these modes all serve to complicate interpretation of test results.

Diametral testing of cores for use in determination of masonry shear strength is mentioned in the City of Los Angeles Building Code [10], among others.

6.4.1 *Equipment*

Most testing agencies should be capable of removing and testing diametral shear specimens using existing equipment. Specimen removal requires use of a water-cooled core drill with a large-diameter diamond bit. Loading of the specimen is accomplished using any type of universal compression testing device.

6.4.2 *Procedure*

The number of core tests is the same as the number of in-place push tests

Figure 6-8. The diametral core shear specimen.

described previously. The desired test location is chosen and the core drill attached securely to the masonry with anchor bolts. Use through-bolts or threaded rods with anchor plates for attachment of core drills to weak masonry. The core diameter should be approximately equal to the length of a single masonry unit for solid clay masonry. A core diameter of 8 inches is adequate for most constructions using modular clay masonry units. Masonry built with larger units requires larger diameter cores. The use of the diametral core shear test on grouted and ungrouted concrete masonry construction has not been reported.

Align the core drill such that a bed joint is centered on the diameter, and the intersection of a head joint and bed joint is at the center of the core as shown in Figure 6-8. The drill operator must exercise extreme care during coring and core removal as the specimen is fragile. Many types of older masonry may not possess sufficient mortar-unit bond strength to allow removal of intact cores; a breakage rate of 50% should be expected for these types of masonry.

The test specimen must consist of a minimum of one wythe of masonry and for most cases a single-wythe specimen is sufficient. If the collar joint is sound and adjacent bed joints are aligned, it may be possible to extract and test multiwythe specimens.

Package the test core carefully for transportation to the laboratory and allow it to dry before testing. Mark the loading axis on the face of the core at an angle of 15 degrees to the diametral bed joint. Parallel loading caps approximately 1/8-inch-thick are made along the two bearing surfaces using a quick-setting gypsum cement.

When the loading caps have cured, the specimen is carefully aligned in a compression testing machine and loaded to failure. The peak compressive load, observed failure mode, and area of test joint are recorded; the average joint shear stress is calculated using Equation 6-7.

Section 8809 of the City of Los Angeles Building Code [10] requires that all test values must be reported, and when cores could not be removed intact a strength of zero was recorded for that specimen. The average of all core tests must be greater than 20 psi, including the zero test results. This imposes a severe penalty for the use of core tests on weak masonry where core removal may be difficult. In these cases it would be desirable to conduct the in-place push test as described above, to avoid the possibility of a large number of zero test values.

6.4.3 *Limitations*

Removal of large-diameter cores from existing masonry buildings is undesirable for obvious aesthetic reasons, and it is often difficult to remove intact cores from weaker masonry. Although the procedure provides a practical means to obtain and test specimens for shear, there are several inherent deficiencies that complicate interpretation of test results. The test is more useful for use as an indicator of shear strength rather than as the basis of a shear strength design value for the masonry.

A large variation in measured shear strength values can be expected: in carefully controlled laboratory conditions an average coefficient of variation of almost 30% was recorded [1]. This large within-test variation requires many samples to obtain confidence in test results.

6.5 *IN-PLACE SHEAR LOAD TESTS*

In many cases the in-place push test and diametral core test do not provide sufficient information regarding masonry shear behavior. The main problem with such small-scale tests is that results of these tests are difficult to extrapolate to overall masonry shear capacity. Masonry piers and shear walls subjected to lateral loadings are subjected to a complicated combination of non-uniform compressive, tensile, and shear stresses, which cannot be adequately modeled by testing of single units and small specimens. Large-scale tests are used when a measure masonry shear strength for a specific building or class of masonry is necessary, or when other techniques are not applicable, such as for determination of shear strength of reinforced or grouted masonry [11].

Many advanced analytical techniques such as the finite element method require accurate information regarding the masonry shear modulus and lateral load-deformation behavior. The shear modulus is difficult to measure in practice and is not easily correlated to results from tests on small-scale specimens. Measurement of deformations during large-scale load tests permits calculation of the masonry shear modulus.

In-place shear load tests are conducted by isolating the element to be tested and simulating seismic and wind loadings by applying lateral loads with hydraulic jacks. The remainder of the structure is used to react to the applied loads. Conducting large-scale tests in-place can be complicated and requires a large amount of planning, labor, and in most cases, repair. An advantage is that the entire wall thickness is loaded, providing an indication of the interaction between individual masonry wythes.

6.5.1 *Equipment*

A section of masonry must be isolated from the remainder of the structure before loading. This requires removal of a portion of masonry on either side of the test specimen as shown in Figure 6-9. A large-diameter track-mounted masonry saw or chain saw with a segmented diamond blade is used to cut through the entire wall thickness.

A hydraulic loading jack with appropriate hoses and pump is used to load the specimen. Dial gages or electronic displacement transducers are

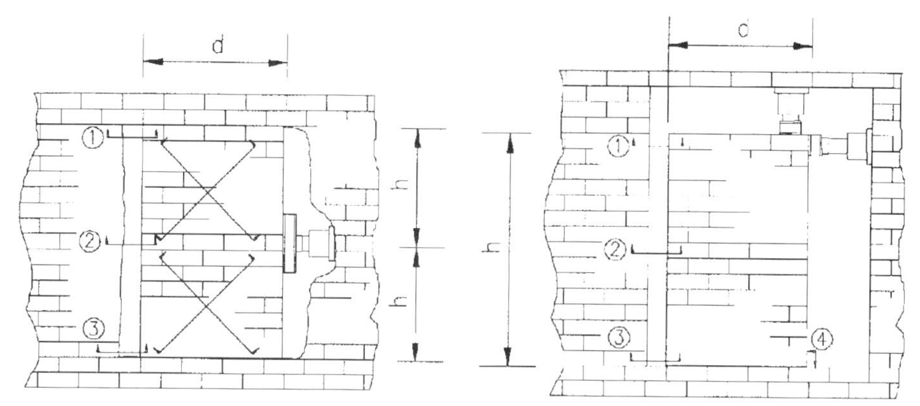

Figure 6-9a. Different techniques for isolating and testing large-scale masonry elements for estimation of shear capacity. Shear test of a pier, fixed on both ends.

Figure 6-9b. An alternate configuration.

attached to the wall for measurement of shear deformations. Deformation measuring devices are placed to record a) overall displacement, b) flexural or rocking deformations, and c) diagonal shear deformations. Deformations should be measured at a minimum of three to five locations as shown in Figure 6-9.

Large-scale in-place testing requires the removal of a significant portion of masonry to isolate the test specimen. Coupled with the possibility of collapse during loading, such a test can result in a potentially hazardous situation. Engineering services should be retained during large-scale testing to evaluate the test configuration and provide recommendations for shoring and support of the adjacent masonry.

6.5.2 *Procedure*

The masonry to be tested is chosen to be representative of typical conditions in the structure, and it is usually preferable to test as small a specimen as possible for equipment and logistics considerations. The specimen must be large enough, however, to provide a realistic response to shear loadings. It is wise to avoid conducting shear tests on critical load-bearing elements to prevent the possibility of structural collapse.

The aspect ratio of the specimen also must be chosen carefully: tall, narrow wall segments may fail in flexure or by rocking; short, squat wall segments are more likely to fail in shear. In general, a specimen consisting of eight to fifteen courses of masonry with an aspect ratio (height-to-length ratio) of approximately 1.0 to 1.5 are acceptable for cantilever specimens. Piers or wall segments fixed at both the top and bottom should have an aspect ratio of two to three. It is difficult to adequately isolate one wythe of masonry from the remainder of the wall, hence it is usually preferable to test the entire wall thickness.

At the chosen test location the specimen is isolated from the surrounding masonry as shown in Figure 6-9. Selecting a pier bounded by window or door openings simplifies this procedure. A vertical section of masonry must be removed from around specimens within walls on either side to allow placement of loading apparatus. This requires saw cutting and removal of several masonry units on either side of the test specimen.

Apply shear loads incrementally until failure occurs. Record deformations for determination of the shear modulus. It may be desirable to include several unloading cycles to obtain information regarding strength and stiffness degradation under repeated loadings. The final failure mode may be dominated by flexural cracking, sliding along bed joints, or diagonal shear cracking. Crack mapping and pictures are useful for describing the final mode of failure.

Repairs will be necessary following testing and in most cases the entire tested portion must be replaced. These repairs can be difficult and will require shoring and bracing until the newly placed masonry has gained sufficient strength to support existing loads.

6.5.3 *Analysis of Test Results*

A plot of the load/displacement curve is useful for determining shear response. Lateral displacement is determined by subtracting the base movement (recorded at locations 1 or 3 in Figure 6-9 (a)) from the overall pier deformation (recorded at location 2 in Figure 6-9 (a)).

Shear strength v_t is calculated by dividing the load at failure P by two times the cross-sectional masonry area A:

$$v_t = \frac{P}{A} \quad (6\text{-}8)$$

where failure may be defined as the peak load or the load that causes major cracking to occur.

The shear modulus is not constant because of non-linear masonry deformations and decreases with increasing lateral displacement. In the initial linear range the shear modulus can be approximated by relating measured horizontal displacements to the shear and compressive deformabilities. If the masonry is considered to be homogeneous and isotropic, then measured displacements can be related to wall axial and shear stiffness. Naturally masonry is neither homogeneous nor isotropic; however, this technique provides a reasonable approximation of the shear modulus. A relationship can be derived from the expression for wall flexibility considering both flexural and shear displacements [12]:

$$\frac{\delta}{P} = \frac{1}{K_o} = \frac{1.2h}{G_oA}\left[1 + \frac{G_o}{E_{33}} \cdot \frac{1}{1.2} \cdot \left(\frac{h}{d}\right)^2\right]$$

(6-9)

where:

δ= measured horizontal deflection
P= applied horizontal force
K_0= initial lateral stiffness
h= wall height
d= wall width
G_0= initial shear modulus
A= wall base area
E_{33}= masonry initial compressive modulus

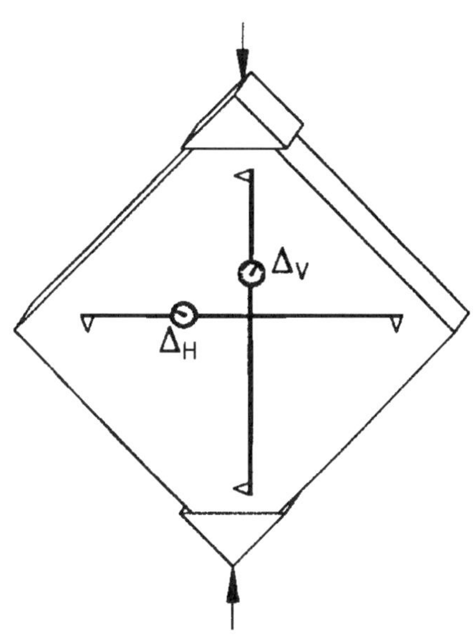

Figure 6-10. Loading and instrumentation for a laboratory test to determine diagonal tensile strength of large masonry panels (ASTM E 519).

If the compressive modulus is known (measured using in-place flatjack measurements or destructive testing) then the shear modulus *G* can be determined using the above relationship.

Shear tests also may be calculated by removing 4-foot-square specimens for laboratory testing as recommended by ASTM E 519, *Standard Test Method for Diagonal Tension (Shear) in Masonry Assemblages*. Instrumentation located along the specimen's diagonals provides information for calculation of the shear modulus described in the Standard (see Figure 6-10). Removal of such large specimens for laboratory testing can be expensive and is usually not desirable. However, the test itself is well-known and the technique can provide a reliable measure of masonry diagonal tension. Failure of these specimens is usually caused by tensile splitting, which provides a lower bound estimate of masonry shear strength.

6.6 CORE TESTS FOR SHEAR BOND

A simple core test is recommended by the Applied Technology Council for determining shear bond strength between grout and the face shell of hollow masonry units. This test does not provide a measure of masonry shear strength as do the previously described techniques; however, it is useful for quality control of grout placement in reinforced masonry construction. It may be possible to identify low-strength grout or improper consolidation by conducting a number of comparative shear bond tests on specimens removed from locations throughout the structure.

6.6.1 *Background*

A description of the core shear bond test is provided in Section 12A of the Applied Technology Council (ATC) "Tentative Provisions for the

Development of Seismic Regulations for Buildings" [13]. The test is proposed as a method for regulating grout materials and grout placement in reinforced masonry. The "Core Test for Shear Bond" (California Test 644) is required by the California Office of the State Architect as a quality assurance test to be conducted on all critical masonry buildings such as schools, fire stations, and hospitals constructed from grouted hollow masonry.

A core sample removed from new masonry is tested to determine bond between grout and masonry. The test is conducted in a compression testing machine by shearing the face shell from the grout. The measured shear stress at failure is used to indicate masonry grout quality.

6.6.2 *Equipment*

A core drill with a diameter equal to about two-thirds of the wall thickness is required. The specimen is removed to the laboratory and tested in shear using a guillotine-type apparatus such as the one shown in Figure

Figure 6-11. The core shear bond specimen uses a guillotine-type of apparatus to determine shear bond between grout and masonry units in unreinforced and reinforced masonry.

6-11. The loading apparatus should be designed to minimize eccentric loads on the bond line, that is, the opposing loads should be applied as close to the joint as possible.

6.6.3 *Procedure*

A single core test is required for each combination of unit and grout type, with a minimum of one test per building or one test per 5,000 square feet of wall area. Cores are taken perpendicular to the wall surface and removed from the center of the grouted cell such that no end or interior webs are included in the plane of the joint. The cell size of most modular hollow-unit masonry limits the maximum core size to a diameter of 4 to 6 inches. Reinforcing steel must be avoided while obtaining core specimens. It is not necessary nor is it advisable, from a repair standpoint, to core entirely through the wall to obtain an appropriately sized specimen. A specimen that includes the unit face shell and 2 to 3 inches of adjacent grout is sufficient.

The core can be removed from the masonry at any time after an age of 14 days; however, it must be saved for testing at an age of 28 days. Cores removed from the wall must be cured in the same manner as masonry prisms, either air or bag cured, depending on the project specifications. Any cores that are broken during removal must be noted.

The cores are tested in a dry condition at an age of 28 days. Measure and record the specimen dimensions, place in the loading apparatus, and load to failure. The peak load is recorded, in addition to the failure mode, for example, failure through the unit, grout, or at the unit-grout interface.

6.6.4 *Analysis of Test Results*

The shear bond strength v_b between grout and masonry units is calculated as:

$$v_b = \frac{P}{\pi d^2/4} \qquad \text{(6-10)}$$

where P is the peak load and d the diameter of the core. The ATC provisions specify that the measured shear strength must be greater than 100 psi.

6.7 *SUMMARY*

Masonry shear strength is an important property for design of retrofits, upgrades, or determination of masonry seismic capacity. Several different techniques with varying levels of complexity have been described for determining masonry shear strength. The techniques described are most useful as indicators of mortar quality throughout the structure and may also be used to quantify long-term deterioration of masonry shear strength or damage to the structure. Correlations with full-scale tests on masonry subjected to lateral loads allows results from the simple tests to be correlated with overall masonry shear strength and can provide an indication of shear strength for use in engineering analyses.

6.8 REFERENCES

1. Noland, J.L., R.H. Atkinson, and J.C. Baur. 1982. An Investigation into Methods of Nondestructive Evaluation of Masonry Structures, National Technical Information Service Report No. PB 82218074. Report to the National Science Foundation.
2. ABK Joint Venture. 1984. *Methodology for Mitigation of Seismic Hazards in Existing Unreinforced Masonry Buildings: The Methodology,* Topical Report 08. Prepared for the National Science Foundation, Contract No. NSF-C-PFR780-19200.
3. International Conference of Building Officials. 1991. *Uniform Code for Building Conservation.* Whittier, Calif.: International Conference of Building Officials.
4. International Conference of Building Officials. 1991. *Uniform Building Code Standard No. 2407, In-Place Masonry Shear Tests.*
5. Schmid, B.L. 1979. Significant Research on Old Unreinforced Masonry Buildings, Parts I and II. *Masonry Industry.*
6. Abrams, D.P. 1992. Strength and Behavior of Unreinforced Masonry Elements. Presented at Tenth World Conference on Earthquake Engineering, Rotterdam.
7. Noland, J.L., R.H. Atkinson, G.R. Kingsley, and M.P. Schuller, Atkinson-Noland & Associates Inc. 1990. Nondestructive Evaluation of Masonry Structures. Report to the National Science Foundation.
8. Atkinson, R.H., G.R. Kingsley, S. Saeb, B. Amadei, and S. Sture. 1988. A Laboratory and In Situ Study of the Shear Strength of Masonry Bed Joints. Proceedings at the 8th International Brick/Block Masonry Conference, Dublin, Ireland.
9. Epperson, G.S., and D.P. Abrams. 1989. *Nondestructive Evaluation of Masonry Buildings,* Advanced Construction Technology Center Document No. 89-26-03. Urbana-Champaign, Ill.: University of Illinois.
10. City of Los Angeles. 1988. *City of Los Angeles Building Code.* Los Angeles.
11. Kato, K., A. Matsumura, T. Endo, T. Kubota, and M. Nishiyama. 1985. Tests on Existing Buildings of Concrete Block Masonry Subjected to Horizontal Forces. Proceedings at the Third North American Masonry Conference, Arlington, Texas.
12. Sheppard, P.F. 1985. In Situ Test of the Shear Strength and Deformability of an 18th Century Stone and Brick Masonry Wall. Proceedings at the 7th International Brick Masonry Conference, Melbourne, Australia.
13. Applied Technology Council. 1978. *Tentative Provisions for the Development of Seismic Regulations for Buildings.* Palo Alto, Calif.: Applied Technology Council.
14. Kariotis, J.C., R.D. Ewing, and A.W. Johnson. 1985. Strength Determination and Shear Failure Modes of Unreinforced Masonry with Low Strength Mortar. Proceedings at the 7th International Brick Masonry Conference, Melbourne, Australia.
15. Abrams, D.P., and G.S. Epperson. 1989. Evaluation of Shear Strength of Unreinforced Brick Walls Based on Nondestructive Measurements. Proceedings at the 5th Canadian Masonry Symposium, Vancouver, British Columbia, Canada.
16. Turner, S. 1992. Evaluation of the Flatjack Test and In Situ Shear Test of Masonry Master's thesis, Georgia Institute of Technology, Atlanta.
17. ASTM. 1988. *Standard Test Method for Diagonal Tension (Shear) in Masonry Assemblages,* ASTM E 519-81. Philadelphia: ASTM.

18. Chiostrini, S., and A. Vignoli. 1994. In Situ Determination of the Strength Properties of Masonry Walls by Destructive Shear and Compression Tests. *Masonry International*. Vol. 7, no. 3.
19. Atkinson, R.H. 1981. Analysis of the Diametral Masonry Core Shear Test Specimen. *The Masonry Society Journal*. Vol. 1, no. 2.

Chapter Seven • *In-place Bond Wrench*

Unreinforced masonry walls subjected to flexural loads can develop bed joint cracks because of tensile failure of mortar-unit bond. Out-of-plane flexural loads can result from eccentrically applied vertical loads, wind loadings, or seismic forces. Because mortar-unit bond of old masonry is typically weak when compared with tensile strength of the mortar and units themselves, mortar bond usually controls the cracking strength of the wall. Determination of mortar-unit bond strength can provide information useful for engineering analysis of wall capacity. Moreover, it can be used to indicate construction quality and material condition throughout the structure.

Tensile cracking in the bed joints does not usually compromise the stability of unreinforced masonry walls because reserve capacity exists due to arching actions even after cracks are first observed. The stress that causes tensile cracks to form is considered to be a serviceability criterion: open cracks resulting from bond failure forms an easy pathway for moisture penetration.

7.1 *BACKGROUND*

Two standard laboratory methods are presently available for determining masonry flexural tensile strength. ASTM Standard Test Method E 518-80, *Standard Test Methods for Flexural Bond Strength of Masonry,* applies to testing of a simple stack-bond masonry prism in flexure. ASTM Method C 1072-86, *Standard Method for Measurement of Masonry Flexural Bond Strength,* utilizes a specially designed bond wrench apparatus to subject a single masonry bed joint to flexural stresses. Both of these techniques provide a measure of flexural tensile bond strength; however, they are not particularly well suited for field evaluations and require removal of test prisms from the structure for laboratory testing.

Shrive [1] adopted the basic concept of ASTM C 1072 and designed a simple clamping apparatus which is readily adaptable to different field applications. This device allows in-place determination of flexural bond strengths in a rapid and cost-effective manner. A comparison study conducted by Shrive has found no significant differences in test results and variability of test values measured using either the field or laboratory devices. A similar device developed by the Building Research Establishment (BRE) in England is portable and light-weight; it incorporates an integral electronic display of applied loads for rapid field evaluations [2].

Bond strength is affected by numerous workmanship and environmental factors and variations in test values can be expected to be between 20% and 40%. Hence a large number of individual tests are required to obtain a reasonable level of confidence in the test results. The field method can be considered to be nondestructive, as the masonry can be returned to its original pre-test condition by replacing the tested units in the wall with new mortar.

7.2 TEST EQUIPMENT

The field bond wrench apparatus consists of a simple clamping apparatus that is fastened to the masonry unit adjacent to the designated test mortar joint (Figure 7-1). Thin sheets of neoprene are used on the bearing surfaces of the clamp to account for slight irregularities of the masonry surface. A 600-foot-pound-capacity torque wrench is attached to the clamp and used to apply a combination of axial and flexural stresses to the test joint. Load is applied manually by a downward force on the end of the wrench extension. The peak applied bending moment is recorded by a slave indicator on the torque wrench dial. A simple calculation determines flexural tensile bond strength of the mortar joint.

The test apparatus may be designed for use on solid masonry, hollow clay or concrete masonry, or hollow clay tile. It may be possible to test grouted masonry if the device can be constructed to deliver a high flexural force. The test is conducted most easily on single-wythe masonry. It

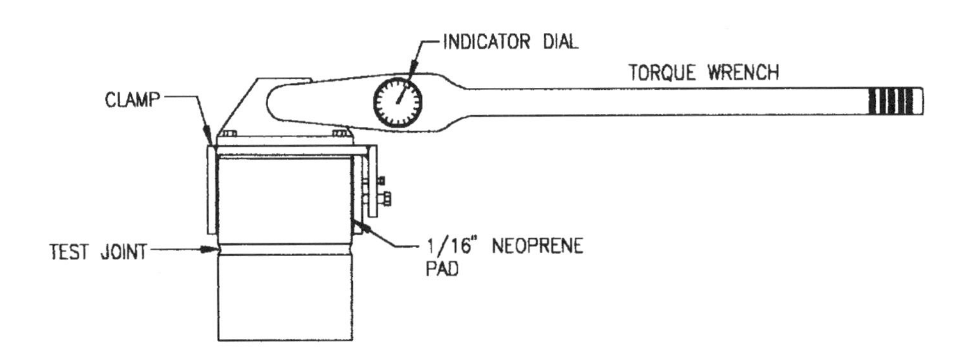

Figure 7-1. In situ bond wrench test apparatus.

would be necessary to remove collar joint mortar on multiwythe masonry to allow placement of the clamping apparatus. A photo of the test apparatus in use is shown in Figure 7-2.

7.3 PROCEDURE

A simplified test procedure for in-place determination of masonry flexural bond strength is as follows:

1. The desired test location is chosen based on engineering objectives, but an attempt should be made to obtain a random sampling of mortar joints throughout the structure in question. The number of tests

Figure 7-2. In-place bond wrench apparatus in use.

should be sufficient to provide a statistically reliable sample (see Chapter 1 for discussion of the number of required tests). A minimum of five to 10 tests is suggested for each area being investigated.

2. Two masonry units on the course adjacent to the test unit are removed, as shown in Figure 7-3, to isolate the test joint and allow a space for attachment of the wrench. The units can be removed by stitch drilling or sawcutting of the mortar joints. Care should be taken to minimize disruption to the test unit during mortar removal. Two units may be tested at this location, one being directly above and the other directly below the removed units.

3. A hydraulic jack is inserted into the opening (Figure 7-3) and a small vertical compressive load, equivalent to approximately 20 psi over

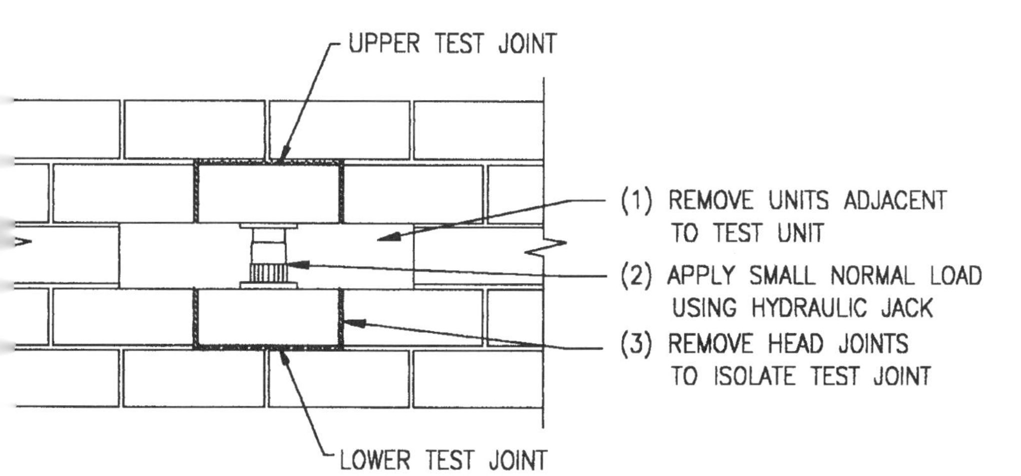

Figure 7-3. Units and head joints removed before test.

the bedded surface, is applied to minimize any vibrational damage to the test joint. The head joints on either side of the test unit are then completely cleared of mortar by stitch drilling, thus isolating the bed joint to be tested.

4. Measure and record the dimensions of the test joint (length, width, and joint thickness).

5. The clamping apparatus is centered horizontally on the test unit and attached using screws. A small torque wrench should be used to adjust the clamping screws to provide the same clamping pressure for every test.

6. A large-capacity torque wrench is attached to the clamp and the indicator dial set to zero. The load is applied normal to the wrench handle in a vertical direction. It is essential that the point of load application be marked on the torque wrench handle and the load applied in the same manner for all tests. The BRE apparatus uses a cross-rod permanently affixed to the wrench for load application.

7. The applied load is increased gradually at a constant rate such that failure of the test joint occurs between 30 and 60 seconds following initial loading. Record the maximum applied torque as indicated by the slave pointer on the indicator dial.

8. Note the type of failure and failure plane, for example, failure at the unit-mortar interface, through the unit, or through the mortar. If bed joint mortar is furrowed or the unit is not fully bedded, estimate the percentage of bedded surface area. A sketch of the bedded area showing mortar placement is useful.

9. Repair the masonry at the test location by replacing any removed units in the wall using a mortar similar in color and composition to the original mortar.

7.4 *ANALYSIS OF TEST RESULTS*

The tested joint is subjected to a combination of flexural loads and either compressive or tensile normal loads, depending on the orientation of the joint relative to the test apparatus. A simple calculation based on linear elastic theory is used to resolve these forces and determine the flexural tensile resistance of the joint at failure.

Applied loads are shown in Figure 7-4. Axial forces on the joint result from the dead weights of the unit, clamping apparatus, and torque wrench, as well as the applied vertical load. These axial forces subject the joint to tension or compression, depending on the orientation of the joint relative to the applied load. Bending moments arise from eccentricities of these vertical loads about the horizontal centroid of the joint. The maximum tensile stress occurring at the test joint is calculated as:

$$f_t = \frac{M_a + P_l\,L_l}{S_j} + \frac{P_l + P_u + P_a}{A_j} \qquad (7\text{-}1)$$

f_t = flexural tensile strength of test joint
M_a = applied moment, from torque wrench indicator
P_l = weight of clamp and torque wrench
P_u = weight of masonry unit
P_a = applied load = M_a/L_a where L_a is the lever arm length
L_l= horizontal distance from the mass centroid of the loading apparatus to the centroid of the tested joint
S_j = section modulus of joint
= $bd^2/6$ for a fully bedded rectangular area, b = average length of joint; d = average width of joint
A_j = area of test joint

The second term in Equation 7-1 is additive for cases where the test joint is located above the test unit (that is, applied load causes additional tensile stress in the test joint) and is subtractive when the test joint is below the test unit (that is, applied load causes compressive stress in the test joint). The individual weights of the clamp and torque wrench must be determined for use in Equation 7-1. The centroid of the loading apparatus is determined by balancing the apparatus on a knife-edge section to determine L_l above.

7.5 *RECOMMENDATIONS FOR DETERMINATION OF ALLOWABLE STRESSES*

Field testing of flexural bond strength is generally conducted for one of two reasons: as qualification testing to verify minimum bond values or to provide a design value for flexural bond.

Engineering analysis for resistance to flexural loads requires a reliable estimation of the flexural bond strength. In cases where archaic

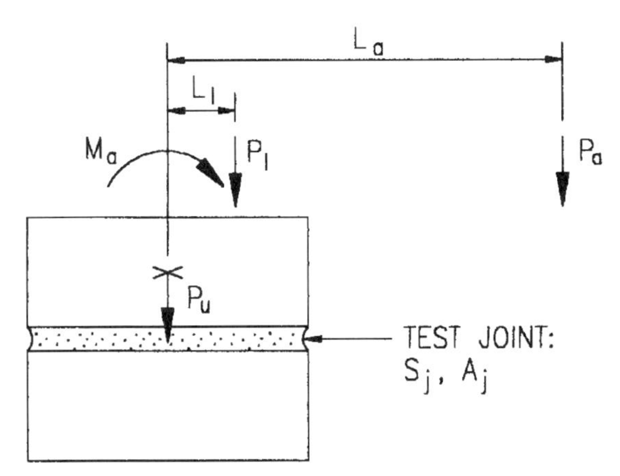

Figure 7-4. Applied loads on test joint.

materials are used, current design codes may not provide appropriate design values. Field test data from in-place bond testing can be used in this situation to determine flexural strength values for masonry.

Following completion of the tests, the data must be analyzed to determine an appropriate design value for flexural bond strength. The mean value alone is not sufficient for determination of the bond limit; some reduced value should be chosen to provide an adequate margin against flexural overload. An appropriate value can be calculated based on the mean and standard deviation of the test results, as well as the required confidence level for strength estimates [3,4]. See Chapter 1 for discussion of statistical methods for interpretation of test results. A 20th percentile value may be used as a conservative measure of material properties for analysis and is recommended for consideration of uncertainties caused by random variations in measured material properties.

Figure 7-5. Field tests for determination of flexural tensile bond were conducted on the exterior walls of this clay tile structure. Note the lack of expansion joints in the wall and the permanent movement of masonry relative to the concrete foundation at corner.

7.6 *CASE STUDY*

A series of in-place bond wrench tests was conducted on the clay tile structure shown in Figure 7-5. The building was constructed in 1942 as a temporary building during World War II and has a floor area of approximately 220,000 square feet with more than 3,000 linear feet of exterior wall. The majority of the building is single-story hollow clay tile construction with a small central portion of the building having two-story masonry walls. All exterior walls are constructed of hollow clay tile that must resist wind, seismic, and roof loads. The clay tile units are nomi-

Figure 7-6. Typical cracks observed in exterior wall. Note movement of masonry relative to foundation wall and step cracking extending diagonally from wall corner.

nally 5 inches high, 8 inches thick, and 12 inches long with horizontal cores, set in a medium-hard cement-lime mortar.

The building was suffering from a prominent system of diagonal and horizontal shear cracks attributed to a lack of thermal expansion joints (see Figure 7-6). In addition, many of the head and bed joints were partially delaminated due to weathering.

A preliminary analysis of the structure was conducted as part of a feasibility study to upgrade the roofing system. This analysis concluded that, according to current design specifications, some portions of the exterior clay tile walls were insufficient for resisting out-of-plane flexural loads. Specifically, design wind loadings resulted in tensile stresses exceeding working stress design values for the masonry. A series of in-place and nondestructive tests was conducted to determine the present condition of the masonry and calculate appropriate material properties for determination of masonry flexural capacity.

7.6.1 *Test Results*

Fifteen in-place bond wrench tests were conducted throughout the structure. Test locations were chosen to be representative of masonry variations, including tests at areas visually classified as being of good, average, and poor quality. Several of the tests were conducted at the base course to determine the condition of bond between the wall and foundation. The bond at the base course was deteriorated from many cycles of thermal movements of the masonry relative to the foundation walls.

Measured flexural tensile bond strength values were relatively low, presumably because of the smooth extruded surfaces of the tiles and deterioration of the joints resulting from thermal movements and weathering. All tested joints failed predominately at the interface between the mortar bed joint and the unit. Some failure within the mortar itself was observed but none

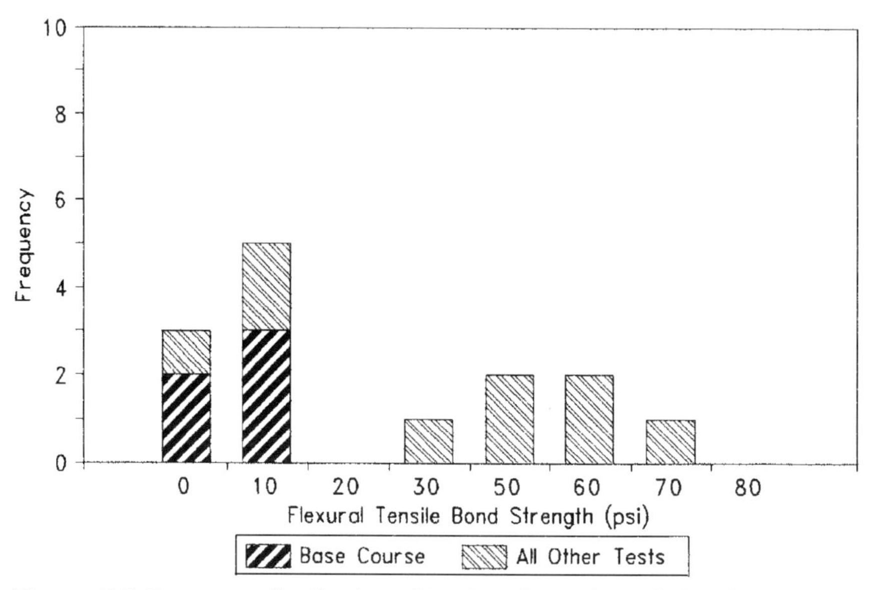

Figure 7-7. Frequency distribution of in-place flexural tensile bond test results.

of the failure surfaces passed through the unit. A distribution plot of bond strength values is shown in Figure 7-7.

Examination of the distribution plot of Figure 7-7 shows two distinct groupings: one consisting of test joints located at the base of the wall, all of which had basically zero bond strength, and another grouping of tests conducted elsewhere within the walls. Significance testing on this data shows there to be a 99.5% probability that the two sample sets are different. Hence these data were split into two separate groupings for determination of statistical data.

The base course tests verified damage to mortar bond resulting from thermal movements. A mean test value of only 0.4 psi was obtained from these specimens, leading to the recommendation that the mortar joint between the masonry walls and foundation be repaired to restore flexural capacity and some degree of fixity at the wall base. The repair to the mortar joints would be made after expansion joints had been installed in the structure.

The mean value for tests conducted at other locations within the walls was 32.2 psi, with a standard deviation of 24.3 psi. A 20th percentile value of 12.8 psi was calculated based on test data, to be used as an allowable stress for analysis purposes. There is little historical information regarding flexural strength of clay tile masonry construction. Test data reported by Plummer [5] has a mean value for flexural tensile stress of 42 psi, with an allowable design value of 10 psi. This compares favorably with test values obtained using the in-place bond wrench apparatus. Results from the in-place testing show that even though flexural tensile bond at the damaged base course was reduced effectively to zero, the remainder of the masonry was relatively undamaged by thermal movements and required no remedial measures.

7.7 *SUMMARY*

In-place bond tests can be conducted inexpensively using simple test apparatus. The field method is considered to be nondestructive, as the masonry can be returned to its original pre-test condition by replacing the tested units in the wall with new mortar. In-place bond tests provide data directly useful to engineering analysis and evaluation without requiring correlations to other material tests. The test may be applicable to other situations and, as experience is gained, application of such a test should be standardized for field applications through the ASTM or other standards organizations.

7.8 REFERENCES

1. Shrive, N.G., and D. Tilleman. 1992. A Simple Apparatus and Methods for Measuring On-site Flexural Bond Strength. Proceedings at the 6th Canadian Masonry Symposium, University of Saskatchewan, Saskatoon, Canada.
2. de Vekey, R.C. 1991. In Situ Tests for Masonry. Proceedings at the 9th International Brick/Block Masonry Conference, Berlin, Germany.
3. Atkinson, R.H. 1992. Statistical Requirements for Masonry Testing, Technical Note. *The Masonry Society Journal*. Vol. 11, No. 1.
4. Natrella, M.G. 1963. *Experimental Statistics*. National Bureau of Standards Handbook.
5. Plummer, H.C. 1992. *Brick and Tile Engineering,* Second Edition. Structural Clay Products Institute.
6. ASTM, 1986. *Standard Method for Measurement of Masonry Flexural Bond Strength,* ASTM C 1072-86. Philadelphia: ASTM.
7. ASTM, 1980. *Standard Test Methods for Flexural Bond Strength of Masonry,* ASTM E 518-80. Philadelphia: ASTM.

Chapter Eight • *Pulse Transmission Techniques*

Pulse transmission techniques are among the most widely used methods for nondestructive evaluation. The test itself is simple in principle, involving measurement of the time needed for an induced stress wave to pass through masonry. Data from this test are acquired rapidly using readily available equipment and can be used for efficient condition surveys. The method is effective at indicating general variations in material condition throughout a structure and also for location of cracks and voids. High-frequency ultrasonic pulse transmission is useful for short path lengths in relatively sound masonry; low-frequency sonic pulses are used for investigations of older or damaged masonry with high attenuative properties. It is possible in some cases to correlate pulse velocity to measured material properties.

8.1 *BACKGROUND*

Pulse transmission techniques are useful for condition surveys and can be used to provide an indication of material variations throughout a large area. Time of flight and velocity measurements are used as indicators of material quality: a low velocity indicates inferior material, deterioration, or damage in the form of cracks and spalls. The technique also can be used on new construction to locate the presence of grout within hollow units and to locate honeycombing, voids, or shrinkage cracks within freshly placed grout.

Pulse velocity does not, by itself, provide an absolute reference to material properties such as compressive strength. However, with companion measurements taken using destructive or in-place techniques, pulse velocity can be correlated to material properties. These correlations normally must be conducted for each structure or combination of mortar and unit type.

The underlying principles of pulse transmission techniques are quite simple. A stress wave is generated in the masonry, usually by an acousto-electric device or a hammer blow. The pulse travels at high speeds through the masonry and is measured at some distance from the generation point and analyzed for velocity, energy content, or frequency content. The stress wave is affected by each material through which it passes and also by each interface between dissimilar materials it encounters. Variations in material soundness and quality can be correlated quite well to pulse velocity.

8.1.1 *Correlation to Material Properties*

Different materials have different pulse velocities, depending on density and dynamic modulus. A dense, stiff material has greater velocity than a soft, porous material. It is this effect which allows correlations to be made between pulse velocity and material quality. The velocity is related to these parameters as follows:

$$v^2 = K\frac{E_d}{\rho} \tag{8-1}$$

where:

v = pulse velocity = $\frac{\text{path length}}{\text{travel time}}$

$K = \frac{(1-\nu)}{(1+\nu)(1-2\nu)}$

E_d = dynamic modulus

ρ = material density

This relationship holds true for homogeneous, isotropic materials and cannot strictly be applied to masonry. The technique does not, therefore, provide an absolute measure of masonry modulus or density.

Pulse velocity may be correlated to masonry material properties by calibration with companion tests. Several studies have been conducted to investigate the validity of relating masonry material properties to pulse velocities, with mixed results. Noland et al. [1] conducted tests on two-wythe clay masonry walls and found there to be a fair correlation between pulse velocity and compressive strength and poor correlations to masonry tension or shear strength. Hobbs [2] was able to distinguish between different grades of masonry and detect variations in both unit and mortar strength using pulse velocity techniques. Calvi [3], on the other hand, found little correlation between pulse velocity and material properties and suggests that a large amount of statistical data would be needed to provide reliable predictions based on ultrasonic testing.

In addition to the types of materials encountered, pulse transmission is affected by each interface between dissimilar materials crossed by the pulse. Interfaces between masonry units, grout, mortar, water, and in the case of cracks or voids, air, all affect the pulse. At each interface a certain amount of the pulse energy is transmitted while a portion of the energy is reflected. Energy transmission is high through interfaces between similar materials such as the face shell of a concrete masonry unit and hardened grout, whereas transmission is somewhat less for boundaries between different materials such as fired clay units and mortar. In the extreme case of an air space, open crack, or void, nearly all of the wave energy is reflected at the interface. Thus a porous or highly fractured material rapidly attenuates the wave energy, resulting in a loss of signal, which severely limits the travel distance of the pulse.

The effect of material quality, voids, and cracks on pulses transmitted directly through multiwythe masonry is shown in Figure 8-1. The test wall shown was constructed with varying materials and construction quality on the interior wythe. Stress waves experience a delay in arrival time when passing through poor-quality materials. An apparent drop in velocity also is observed where large voids interrupt the direct path length. An increase in path length results in an apparent drop in velocity as the wave must travel around the void to reach the receiver. It is also interesting to note the effect of varying materials on the energy and frequency content of the pulse, as indicated by the amplitude and spacing of "peaks" of the recorded pulses.

The ability of pulse transmission techniques to identify variations in material properties and features such as cracks and voids is mainly

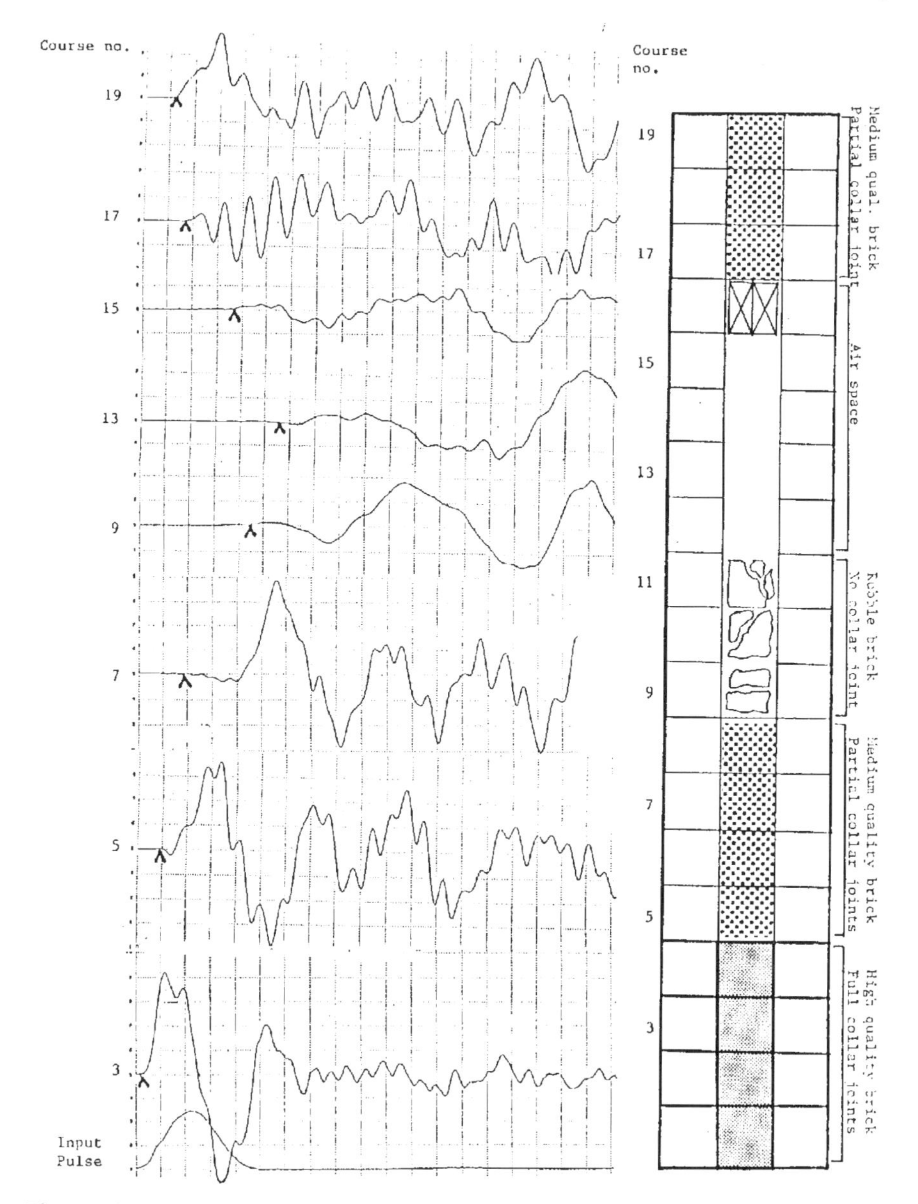

Figure 8-1. Variations in wave forms caused by passage through different materials. The transmitted pulse is shown at the bottom of the figure.

dependent on the dominant frequency and energy content of the transmitted pulse. The input frequency is determined by the mechanical characteristics of the transmission apparatus. Most transmitters for concrete testing operate in the ultrasonic range and generally have an input frequency of about 55 kilohertz (kHz). Such high-frequency signals are useful for locating relatively small voids and cracks but are prone to rapid attenuation of waveform energy. Wave attenuation is more prevalent in older or fractured masonry and the wave often loses all measurable energy after passing through only 2 to 3 feet for such materials. A more robust low-frequency signal in the range of 1 to 10 kHz with a greater energy content generated by a mechanical pulse hammer blow is used for evaluations of larger expanses of old masonry.

8.1.2 *Ultrasonic Pulse Transmission*

Ultrasonic techniques for masonry evaluation use high-frequency pulses in the range of 20 to 100 kHz. Ultrasonic testing has been used successfully for examination of concrete structures with some degree of correlation to concrete compressive strength. However, the high frequency stress wave used for ultrasonic testing attenuates rapidly in most masonry materials. Ultrasonic pulse techniques are most useful on modern, relatively sound masonry, with path lengths of 10 to 30 inches. Low-frequency pulses are more resistant to signal attenuation and sometimes are more useful for masonry testing.

8.1.3 *Sonic Pulse Transmission*

Sonic pulse techniques use an input signal with a frequency ranging from 1 to 10 kHz. The technique often is referred to as "mechanical pulse" testing when the input signal is generated by a blow from an instrumented hammer. The hammer is precisely manufactured and fitted with an internal accelerometer for measurement of the input pulse. Interchangeable heads for different input pulses are available. The screw-on heads have a variation in hardness, ranging from soft rubber to hard rubber, aluminum, or steel tips for producing input pulses with different frequencies. Softer tips produce a low-frequency pulse and harder metal tips produce a higher-frequency pulse. A small metal plate fixed to the masonry surface is sometimes used when using metal-tipped hammers on softer masonry to avoid damage to the brick exterior.

The main advantage of using sonic pulse transmissions is the high energy content of the transmitted stress wave. A 3-pound hammer is useful for most masonry investigations; a pulse generated by a firm tap from the hammer can be recorded up to 20 feet away for global measurements of masonry quality. A 12-pound sledge has been used for generation of high-energy pulses for transmission through more than 30 feet of masonry in massive bridge abutments.

Several research investigations have determined that mechanical pulse testing shows promise for masonry evaluation. Studies by Noland [1,4] concentrated on evaluation of old, unreinforced masonry. Mechanical pulse testing was useful for location of larger cracks and voids in test walls constructed in the laboratory. The technique was more useful than ultrasonic testing in most cases because it was affected by only significant flaws. Relationships between mechanical pulse velocity and compressive strength show a positive degree of correlation but severe data scatter may limit this application.

TABLE 1. TYPICAL PULSE VELOCITIES FOR MASONRY MATERIALS (SOME VALUES FROM [5])

Material	Average Transmission Velocity	
	(m/s)	(fps)
Good brickwork (Uncracked)	3,100	10,000
Poor brickwork (Cracked)	2,500-2,700	8,000-8,900
Uncracked reinforced cavity	3,500	11,500
Cracked reinforced cavity	2,700-3,000	8,900-9,850
Grouted concrete masonry	2,750-3,650	9,000-12,000
Old masonry, weak mortar (Horizontal velocity)	460-2,500	1,500-8,000
Old masonry, weak mortar (Vertical velocity)	365-2,100	1,200-7,000
Hollow clay tile—good quality	1,500-3,000	5,000-10,000
Hollow clay tile—poor quality	914-2,100	3,000-7,000
Structural concrete	4,200-5,200	14,000-17,000
Steel	5,800	19,000

Note: Deteriorated masonry has varying velocities, depending on the number and size of cracks or voids present.

Komelyi-Birjandi [5] used low frequency (1-2 kHz) mechanical pulse transmissions for evaluation of shear damaged masonry. Sonic pulse velocity measurements showed a good correlation to masonry quality. Typical pulse velocity measurements obtained in this study for reinforced clay masonry are shown in Table 8-1. Additional pulse velocities for grouted concrete masonry, older brick masonry, and hollow clay tile construction also are provided in Table 8-1.

Forde [6] used sonic pulse velocity measurements for location of voids in massive masonry bridge abutments. High-amplitude sonic pulses were the only type of stress wave able to penetrate fully through the 9-meter-thick abutment.

Carino [7] used a form of low-frequency testing to conduct impact-echo tests on reinforced concrete specimens. The pulse in this case was generated by a metal ball striking the surface to provide a distinct and repeatable input pulse. The impact-echo technique locates the receiver directly adjacent to the input point and records reflections of the stress wave from internal features. This technique would be difficult to apply to many older masonry structures because of their highly variable material nature; however, it may be applicable to testing of modern reinforced masonry. Impact-echo may be useful for locating grouted cells in reinforced masonry or determining the quality of interior collar joints. Equipment for impact-echo testing is manufactured commercially for concrete evaluation.

8.1.4 *Location of Voids and Cracks*

Both ultrasonic and sonic pulse techniques can be used to locate voids and cracks on the masonry interior. Ultrasonic stress waves are more sensitive to location of smaller voids but are prone to rapid signal attenuation and may not be useful on highly deteriorated masonry. For these cases, sonic pulse testing is preferred. There is a trade-off with using low-frequency pulses: the longer wavelength of sonic pulses causes many smaller voids and anomalies become "invisible" to the wave. Wavelength λ is related to frequency f and pulse velocity v:

$$\lambda = \frac{v}{f} \qquad (8\text{-}2)$$

Voids significantly smaller than the wavelength do not affect pulse transmission and hence are not identifiable. Using a 30 kHz pulse, Leeper [8] found velocity measurements were successful for location of 4- to 6-inch-diameter objects in grouted masonry walls. By examining attenuation characteristics such as energy loss, rise time, and amplitude of the received signal, smaller defects could be located.

8.1.5 *Effect of Reinforcing Bars*

Velocity measurements taken in the vicinity of reinforcing bars may provide inconsistent and highly variable results. The velocity of stress waves through steel is much greater than the average velocity through masonry (see Table 8-1). If the bars are perpendicular to the advancing wave front as shown in Figure 8-2, the overall change in velocity will usually be negligible given the average variability of pulse velocities in masonry. If, on the other hand, the bars run in the same direction as the transmitted pulse (Figure 8-2), the wave will travel much more quickly along the bar, resulting in an apparent increase in pulse velocity, which is not representative of the masonry at that location.

8.1.6 *Test Standards for Ultrasonic and Sonic Testing of Masonry*

Currently, no test standard exists for pulse transmission techniques as applied to masonry evaluation. The European agency RILEM is preparing draft standards for both ultrasonic and sonic testing of masonry. The ASTM

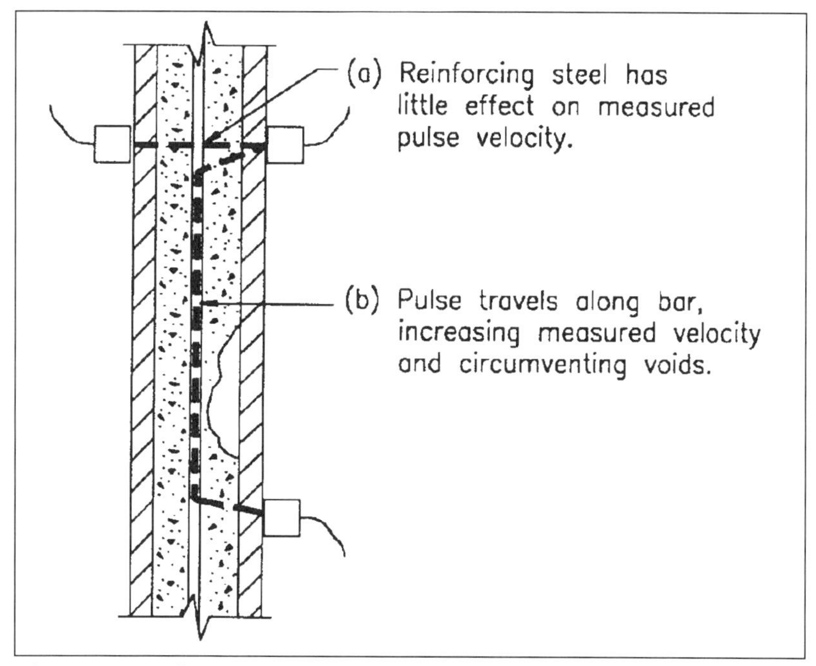

Figure 8-2. Effect of steel reinforcing bars on pulse transmission.

C 597, *Standard Test Method for Pulse Velocity Through Concrete,* describes ultrasonic pulse velocity testing for hardened concrete. The basic procedure described in the standard would be the same for masonry evaluation.

8.2 ULTRASONIC PULSE VELOCITY

Several manufacturers produce commercial equipment for determining ultrasonic pulse velocity. The systems are all portable and may be operated by one or two people. An ultrasonic testing package consists of the following:

- Two transducers, one for signal transmission, the other acting as a receiver
- Main processing unit
- Electronic cables
- Calibration bar
- Couplant

A photo of one type of ultrasonic pulse velocity equipment is shown in Figure 8-3.

8.2.1 Transducers

A small piezoelectric crystal forms the heart of an ultrasonic transducer. When a voltage is applied to the crystal, electrical energy is transformed to mechanical energy and the crystal resonates at a particular frequency. Different transducers can be purchased, each design transmitting in a specific frequency range. Normal transducers for ultrasonic testing of masonry are about 2 inches in diameter and transmit at a frequency of 50 to 100 kHz. This energy is transferred to the masonry when the transmitter is coupled with the material surface. After the pulse travels through the masonry, it is picked up by the receiver. In a similar manner, the piezoelectric crystal in the receiver transforms the mechanical energy of the stress wave to electrical energy and sends this signal to the processing unit.

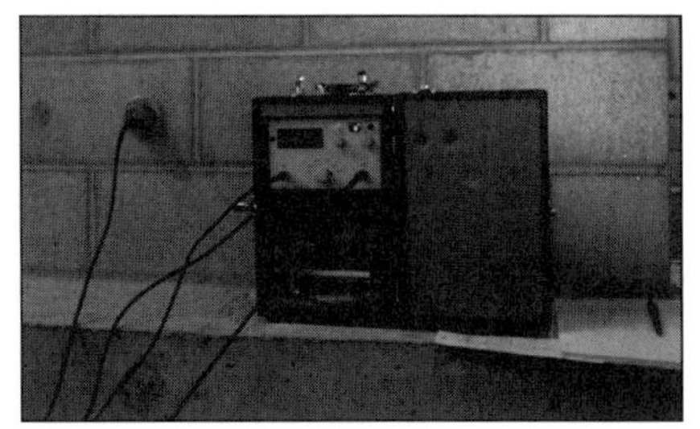

Figure 8-3. One type of ultrasonic pulse velocity equipment.

8.2.2 *Processor*

The main power unit drives the transmitting transducer, supplying power for approximately 10 pulses per second. The receiving transducer sends its signal to the processor. A high-speed internal clock determines the time elapsed between sending the signal and receipt of the first waveform by the receiver. The elapsed time is displayed on the front panel.

The processing unit may contain internal batteries for use in areas where electrical power is not available. Ultrasonic testing systems are usually portable and weigh less than 20 pounds.

8.2.3 *Calibration*

The time recorded by the processor is affected by the type of transducer, environmental factors, and the length of cable used for transducer attachment. Most ultrasonic testing units are supplied with a metal bar for calibration of the processor. The pulse travel time for the bar, determined in the laboratory, is stamped on the side of the bar. Before use, the processor must be adjusted to read the correct transit time on the calibration bar. This process is simple and uses an adjustment knob on the front panel of the power unit.

8.2.4 *Couplant*

A couplant is used to provide contact between the transducer face and masonry surface. Air is a poor medium for transfer of acoustic energy, and on even the smoothest surfaces it is necessary to use a couplant. Many materials are acceptable for use, but all couplants must satisfy several basic requirements. An acceptable couplant must provide maximum transfer of pulse energy from the transducer to the masonry, must not significantly affect transit time, and if possible should not mar or permanently stain the masonry surface.

On smooth surfaces a thin, pliable rubber disk may be used to couple the transducer to the masonry. Most masonry surfaces, however, require a paste or fluid mixture to displace all air from surface depressions and pores. Heavy grease, putty, wax, and commercially manufactured gels all work well as couplants.

Couplants should be applied at the chosen location as sparingly as possible, as thick coatings affect pulse transmission. A thickness up to 0.05 inch is normally acceptable and has a negligible effect on measured pulse velocity. It is important to fill all surface irregularities with the couplant. Extremely rough surfaces require grinding or filling with plaster to provide an acceptable contact surface.

A stable reading that does not fluctuate by more than 1% indicates adequate coupling. Large fluctuations in arrival time with increasing pressure on the transducer indicate inadequate coupling.

8.2.5 *Determining Pulse Velocity*

The transducers must be precisely positioned before determining pulse velocity. The location of both the transmitter and receiver should be well marked and the distance between them measured to within 1%. Couplant is applied and the transducers pressed firmly against the surface. Two people will be required if the path length is long or if the transducers are located on opposite wall surfaces. The elapsed time readout should stabilize rapidly, as indicated on the processor. When the signal becomes weak, due to passage through highly fractured masonry or with a long path length, the processor may not be able to identify the initial wave front and the transmit time will fluctuate severely. In such cases, report the fluctuation and move the transducers closer or to another location until a stable reading can be obtained. Following a successful velocity reading, the transit time is recorded and the transducers are moved to the next position. Rapid scanning of large areas can be accomplished using ultrasonic testing, usually taking less than one or two minutes for each reading using two people. Detailed information regarding types of tests, positioning of the transducers, and data applications is discussed in later sections.

8.3 *MECHANICAL PULSE VELOCITY*

Currently, no commercial manufacturers make low-frequency sonic test system for masonry. All necessary components for producing and recording wave forms are available, but no main processing unit has been developed. A mechanical pulse system relies on the equipment operator to identify the transmit time of the recorded waveform. Equipment necessary for low-frequency mechanical pulse testing, shown in Figure 8-4, includes the following:

- Instrumented hammer for pulse generation
- Accelerometer for receiving the waveform
- Two accelerometer power supplies
- Waveform recording device such as an oscilloscope or a digital waveform recorder for measurement of travel times
- Portable computer system or printer, if records of the pulses are required

8.3.1 *Pulse Generation*

Hammer kits are available in several standard sizes with heads of varying hardness and weight for generation of a wide range of input pulses. Pulses

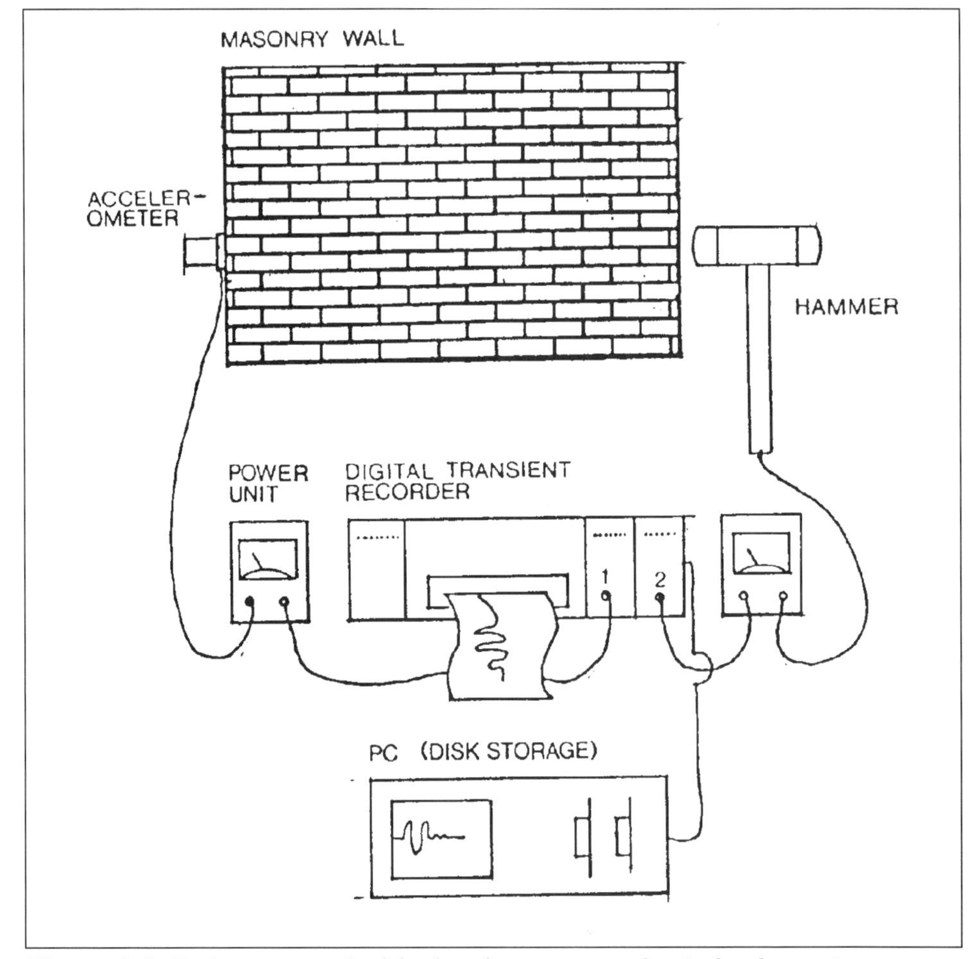

Figure 8-4. Equipment required for low-frequency mechanical pulse testing.

with frequencies ranging from 1 kHz to 10 kHz and capable of long transmission lengths, through 30 feet of masonry and more, are produced using different types of rubber or metal heads. Softer rubber heads may be used on sensitive surfaces to avoid damage. Documentation provided by the manufacturer identifies performance characteristics of the hammer.

A small accelerometer is located within the head of the hammer. This transducer operates on the same principle as the ultrasonic transducers described above, transforming mechanical energy to electrical voltage using a piezoelectric crystal. The voltage signal produced during a hammer hit is used as trigger, causing the oscilloscope or waveform recorder to begin acquiring data.

8.3.2 *Accelerometer*

An additional accelerometer is mounted on the masonry surface at the opposite end of the transit path. An accelerometer sensitivity of between

100 and 1,000 mV/g is adequate for most applications. The accelerometer must be coupled to the masonry surface for adequate pulse transmission using a wax, heavy grease, or commercially produced couplants. If testing personnel is limited it may be preferable to attach the accelerometer to the masonry using a quick-setting adhesive. Low-bond hot-melt glue works well for these applications and is easily removed from the masonry following the test.

8.3.3 *Power Supplies*

Most accelerometers require a DC power supply for operation. Small portable power supplies incorporating batteries and internal amplifiers can be purchased commercially. The power supply also receives the electrical input from both the hammer and receiver accelerometers, amplifies and sends this signal to the waveform recorder.

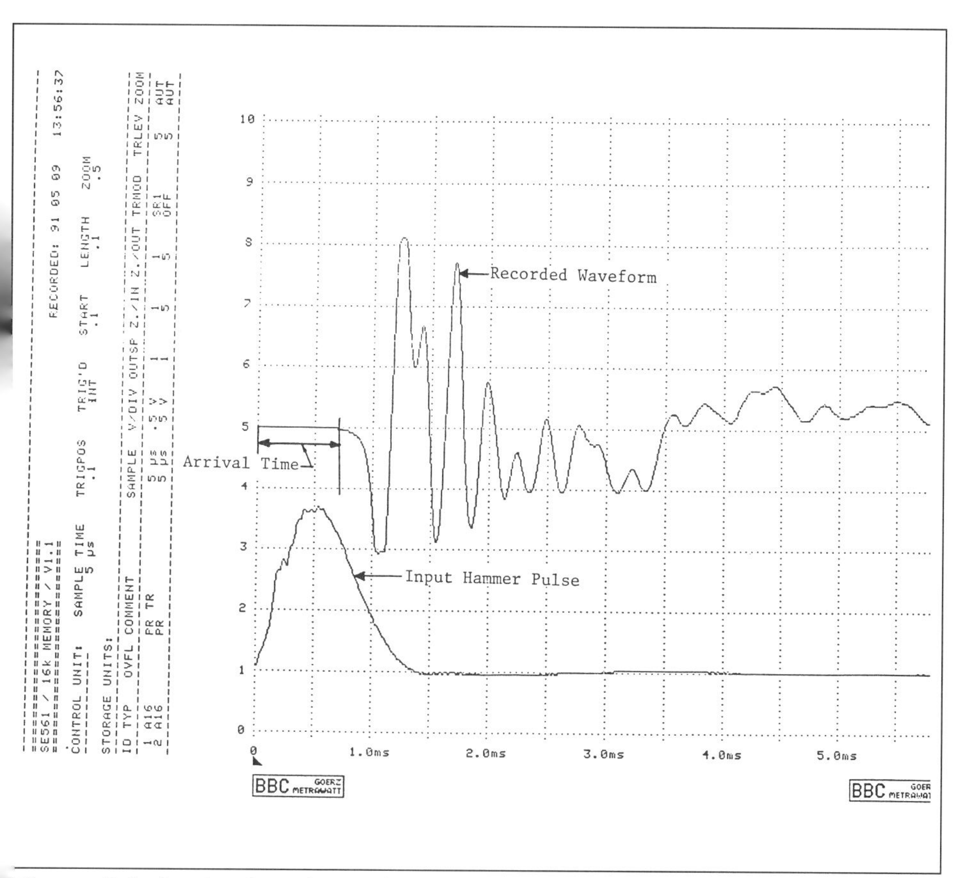

Figure 8-5. A pair of mechanical pulse wave forms. The bottom trace is the transmitted pulse; the upper trace shows the received waveform. Pulse arrival time is determined by the time elapsed between onset of the hammer pulse and the first arrival of the recorded waveform.

8.3.4 *Waveform Recorder*

Because no processors are available for sonic masonry testing, it is necessary to examine each pulse for determination of pulse transmit time. If an experienced operator is available, this can be accomplished on-site using a high-resolution dual-channel oscilloscope. When the masonry is struck with the hammer, the voltage measured by the internal accelerometer (Channel 1) is used to trigger the recorder to begin recording data. The pulse recorded by the receiving accelerometer (Channel 2) is shown on the same display and the elapsed time between the initial rise of the hammer strike and the initial wave front recorded by the receiver is determined as the pulse transit time. A pair of sonic pulse traces showing both the hammer hit and received waveform recorded with a digital recorder is shown in Figure 8-5.

High-resolution oscilloscopes and digital waveform recorders also may incorporate internal printers or interfaces for storage using portable computers. These options allow later analysis and interpretation of the wave using more sophisticated frequency analysis techniques.

8.3.5 *Obtaining a Reading*

Test positions are marked on the masonry surface and the receiving accelerometer mounted at the pulse pickup point; the input pulse is generated by a hammer blow at the marked input point. Several repetitions may be necessary to obtain a hammer hit with the required amplitude. Transmit time is not affected by minor variations in input energy, but the hammer operator must strive to produce consistent pulses. For more sophisticated analyses where energy content of the input pulse must be controlled, an optional pendulum-type device may be fabricated to provide precise input pulses [9].

The digital waveform recorder is set to trigger on the hammer pulse. When the operator strikes the masonry with the hammer, the recorder is triggered to begin acquiring data from the hammer and the receiver. The time elapsed between onset of the hammer hit and the initial waveform recorded by the receiver, as shown in Figure 8-5, is recorded as the pulse transit time.

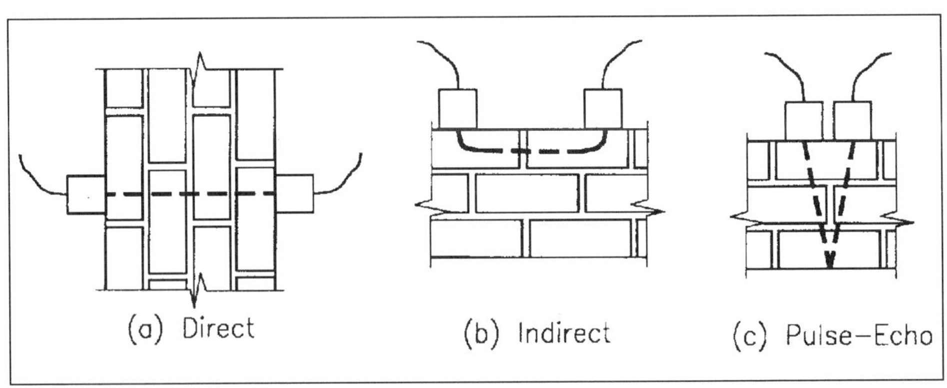

Figure 8-6. Transducer orientation for pulse transmission.

8.4 *TEST PROCEDURE*

Many types of evaluations can be conducted using pulse transmission techniques. Pulse velocity can be used to determine general material condition, locate subsurface cracks and voids, and indicate variations in material properties.

Pulse transmission techniques are most useful for initial investigations where a general survey is conducted with large transducer spacings to locate areas that may be flawed. Following the initial survey, more detailed pulse testing is conducted on potentially deficient areas to determine the nature and extent of the anomaly. If the detailed scan shows significant deficiencies, the area should be checked further using in-place tests or destructive probing.

Pulse velocity readings normally are taken using one of two basic transducer orientations: direct or indirect, as shown in Figure 8-6. Pulse-echo techniques also may be useful. Pulse velocity for all cases is calculated simply as path length divided by travel time and is normally expressed in feet or meters per second.

8.4.1 *Direct Transmission*

The direct, or through-wall, technique is preferred if access to both sides of the masonry is available. This transducer orientation best uses the pulse energy and allows investigation of interior wythes, collar joints, and grouted cavities.

A dense gridwork of pulse velocity readings taken over a wall surface using the direct technique is useful for locating internal voids and indicating general material quality. Results of direct readings are tabulated for all pulse velocities for a general area. Pulse velocities significantly less than the mean value indicate deficient areas. An alternative method for data presentation is to construct a contour plot of pulse arrival time as shown in Figure 8-7. Areas with smaller arrival time (better material quality) appear as valleys and depressions, and areas with high transit times (poor material quality) are immediately evident as peaks and ridges. The contour plot of Figure 8-7 displays the presence of a large void in the interior masonry wythe.

8.4.2 *Indirect Transmission*

Indirect readings also are useful but more prone to rapid signal attenuation than direct readings. The pulses leaving the transducer are propagated most strongly normal to the face of the transducer, and only a small portion of the energy radiates outward along the masonry surface. The indirect technique indicates general variations in general material condition and also locates cracks oriented perpendicular to the masonry surface. A disadvantage of the indirect method is that only the outer wythe or layer of masonry is investigated.

Indirect readings are best obtained by affixing the transmitter to the masonry at one location and taking several subsequent readings while moving the receiver progressively farther from the transmitter. These readings

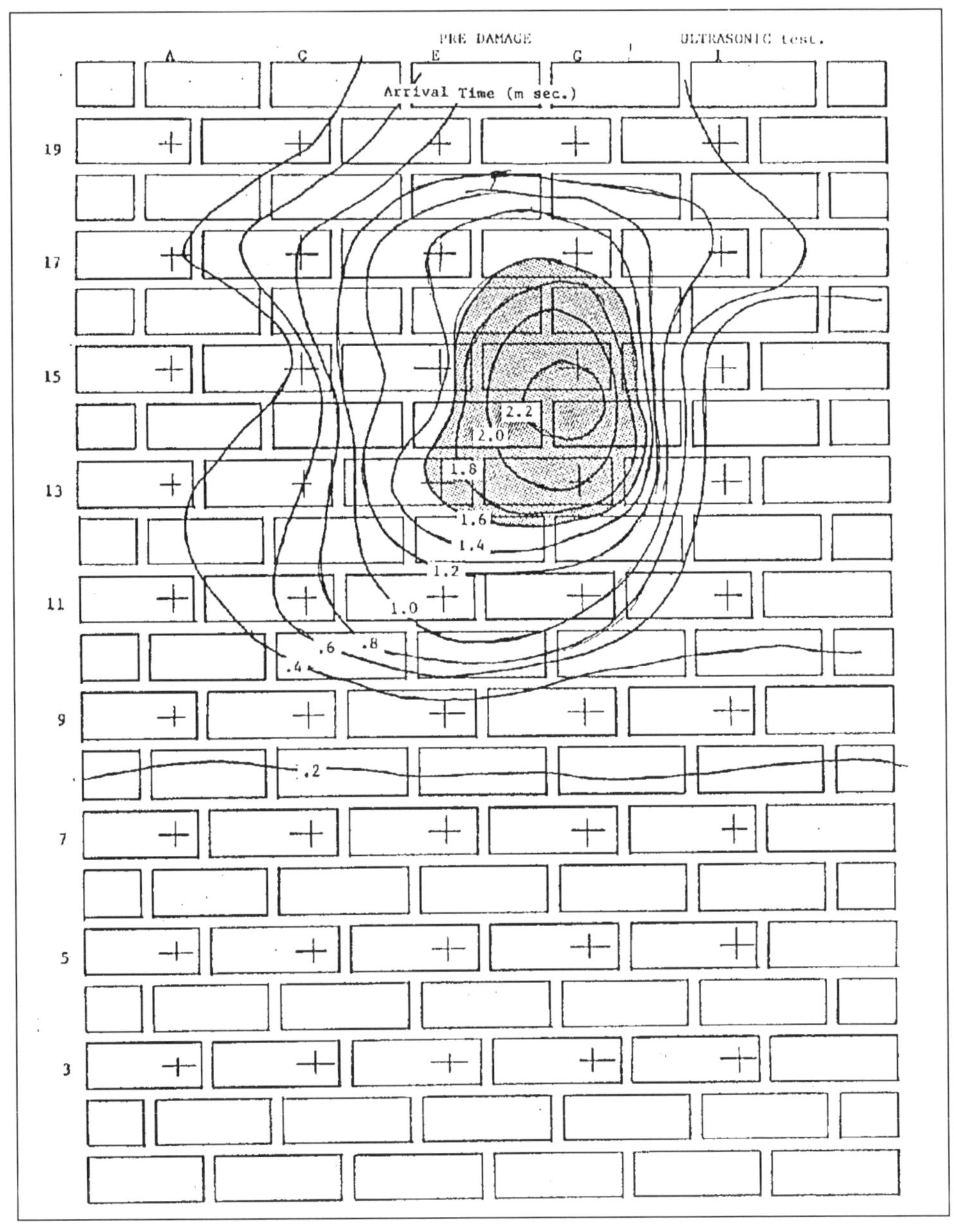

Figure 8-7. Contour plot of through-wall pulse arrival time. A large internal void is evident.

may be taken along either a horizontal or vertical line on the masonry surface. Vertical pulse velocities normally are less than horizontal pulse velocities because of the closer spacing of masonry joints in the vertical direction.

A plot of path length versus arrival time indicates material uniformity along the path. The slope of a line connecting the points gives the aver-

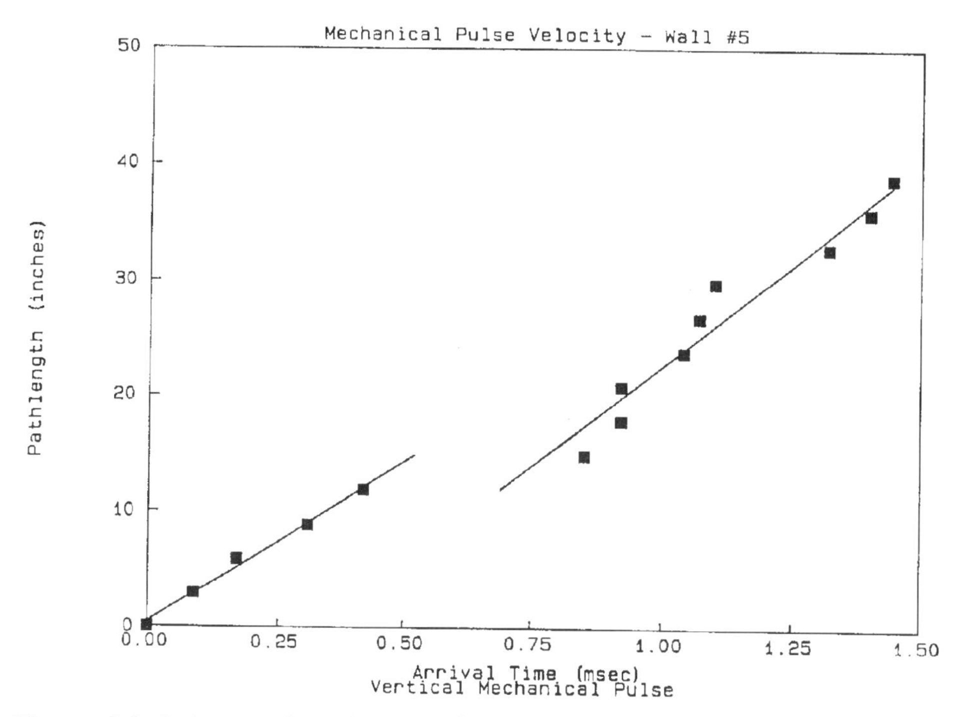

Figure 8-8. Indirect pulse velocity readings taken along the surface of a masonry wall. A crack at about 15 inches causes a noticeable break in the curve.

age pulse velocity along the path. A break in the velocity profile indicates a single crack between receiver positions, as indicated at a path length of about 15 inches in Figure 8-8. A complete change in slope, on the other hand, indicates a transition to highly fractured materials or masonry having significantly different material properties. This is shown in Figure 8-9 where at a pulse length of about 30 inches pulse velocity is reduced.

In some cases pulse velocity alone does not provide sufficient information. Pulse velocity may not be affected significantly by some types of flaws such as numerous small voids or cracks. Remember, however, that as the pulse passes through each interface, a portion of the incident waveform energy is reflected. Higher frequency components are more prone to energy loss and attenuation; hence, a waveform passing through many flaws is dominated by low-frequency components.

This effect is shown in Figure 8-1 where waves traveling through deficient masonry lost much of the energy and higher frequency components of the initial signal. Hence in many cases analysis of the received waveform for energy and frequency content provides important information. This analysis can be conducted in the frequency domain using Fourier transforms and signal processing software. Alternatively the operator can qualitatively grade the received waveform based on the shape and strength of the wave in the time domain. An analysis of this type is described later in this chapter.

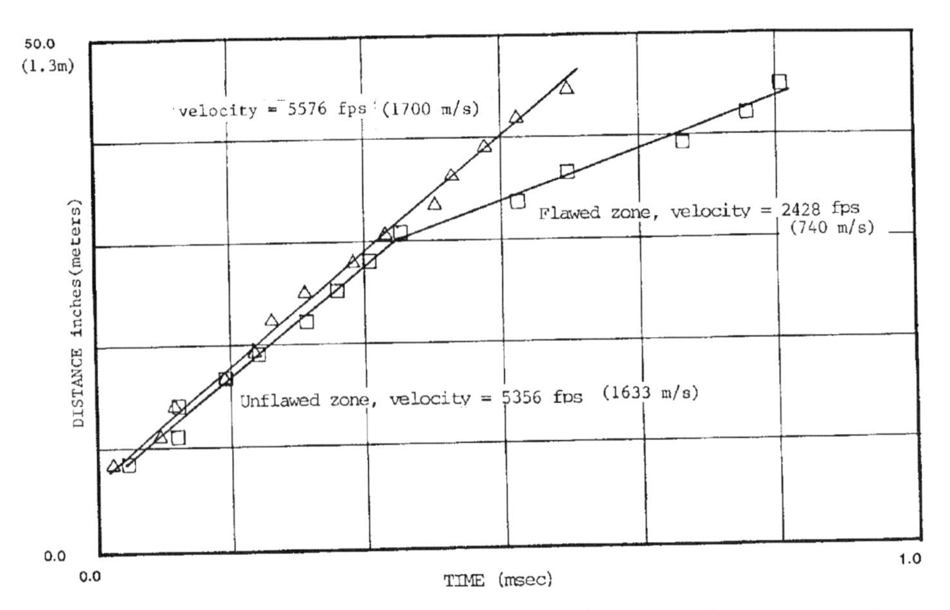

Figure 8-9. A series of indirect pulse velocity readings. Two distinct material conditions are evident: passage into deteriorated masonry at about 30 inches results in a decrease in pulse velocity.

8.4.3 *Labor Requirements*

A major advantage of pulse velocity testing is that large areas of masonry may be evaluated in a rapid manner. Conducting several horizontal and vertical indirect readings on a large masonry panel will usually require less than two man hours. More detailed investigations require additional time: expect to spend 8 to 12 man hours to lay out a measured grid pattern on a 4-foot-square wall panel and record through-wall velocity readings at 4 inches on center.

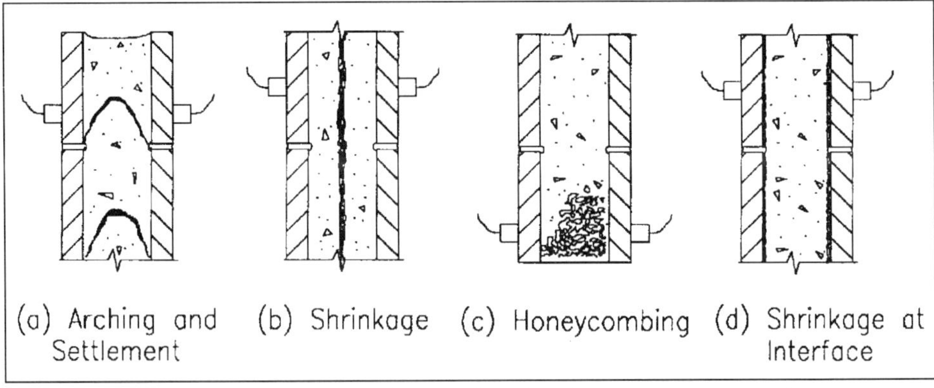

Figure 8-10. Grout shrinkage and consolidation flaws can be located using direct transmission techniques.

8.5 APPLICATIONS

Pulse transmission techniques can be used for many application where knowledge of variations in material quality, locations of voids, or determination of material condition is required. Several examples of typical applications are described below.

8.5.1 Quality Control during Construction

Pulse transmission techniques can be used to monitor workmanship and material qualities during construction. Improper construction procedures, including furrowing of mortar bed joints, unfilled head and collar joints, voids in grout, and improper mortar placement around flashing and shelf angles can be identified using pulse transmission techniques. Ultrasonic pulse testing would be acceptable for most modern masonry

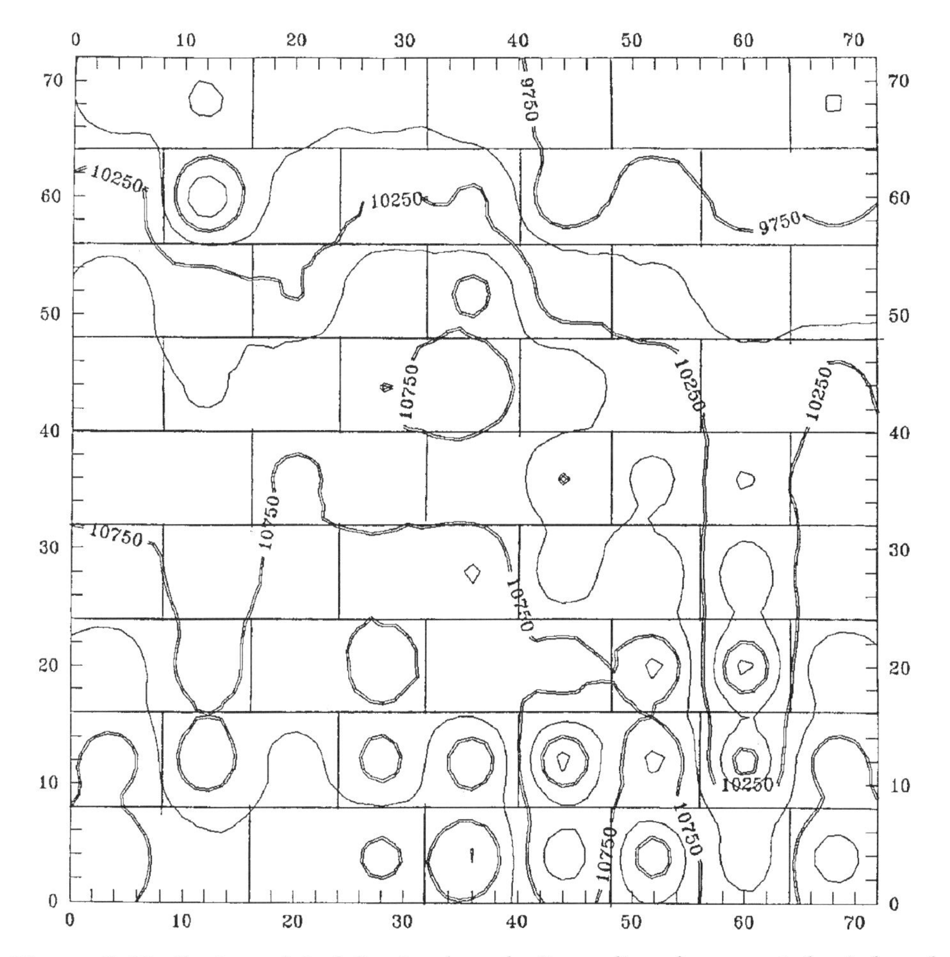

Figure 8-11. Contour plot of direct pulse velocity readings for a grouted, reinforced masonry wall. Note the decrease in pulse velocity toward the top of the wall and poorly consolidated cell near the right edge.

Figure 8-12. The Benjamin Powell Smokehouse, Colonial Williamsburg, Va.

when solid or grouted units are used and where the path length is less than about 3 feet. It is important to take into account the age of the masonry for new construction: Hobbs [10] discovered that ultrasonic pulse velocity will increase for a period of approximately three days, after which the velocity stabilizes to a constant value.

8.5.2 *Testing of Reinforced Masonry*

Modern reinforced masonry can be evaluated using pulse transmission techniques. When testing reinforced masonry it is important to first locate all reinforcing bars. As discussed previously, the presence of reinforcement affects pulse transmission and must be taken into account.

Grout shrinkage cracks and consolidation flaws shown in Figure 8-10 can be identified using simple velocity measurements. Senbu et al. [11] was able to locate grout flaws using ultrasonic pulse velocity and also found a correlation between unit absorption properties and pulse velocity for unflawed specimens.

The contour plot of through-wall ultrasonic pulse velocity shown in Figure 8-11 shows variations in grout quality on the interior of a grouted, reinforced masonry wall. A decrease in pulse velocity near the right edge of the wall indicates that, although there is grout in this cell, it is of lower quality than grout in adjacent cells. It is likely that this cell was not properly consolidated, leaving many voids and/or plastic shrinkage cracks. Another interesting observation is that there is a general trend of decreasing pulse velocity toward the top of the wall. This effect has been reported for reinforced concrete construction and must be considered when evaluating pulse velocity results.

8.5.3 *Interior Masonry Condition*

The Benjamin Powell Smokehouse was constructed in the early 1800s and is located in Colonial Williamsburg, Va. The smokehouse, shown in Figure 8-12, is constructed of hand-molded brick masonry with 16-inch-thick walls. Surface deterioration, presumably caused by salt crystallization and weathering, has led to rapid decay of the historic structure.

Exploratory through-wall tests on the highly variable masonry walls showed that ultrasonic signals attenuated rapidly and would not effectively penetrate the full wall thickness. Low-frequency, high-amplitude

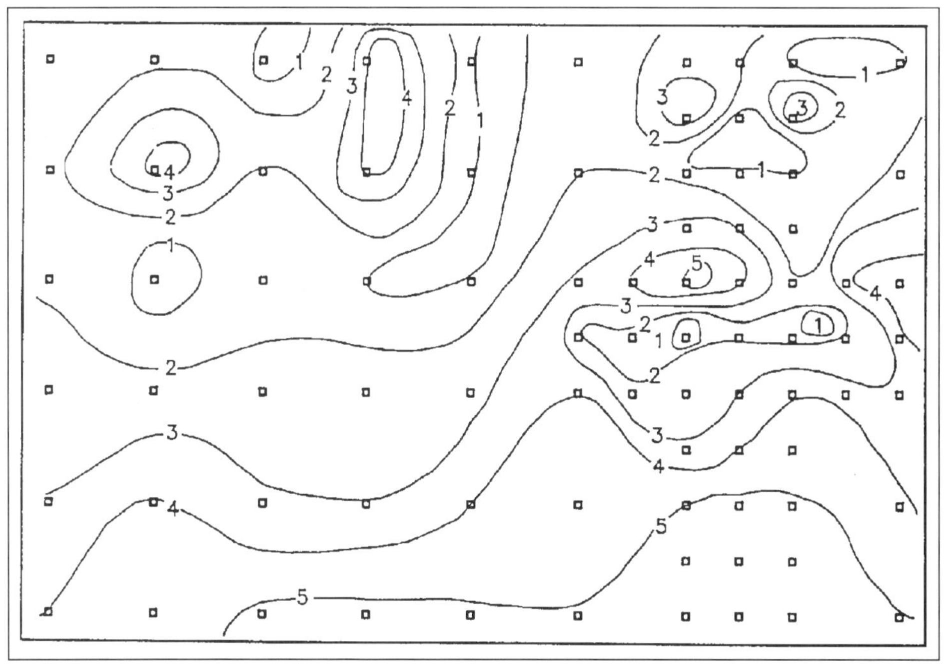

Figure 8-13. Contour plot of through-wall pulse transmission quality. Level 1 readings indicate a high-quality pulse and sound masonry. Level 5 readings represent a waveform which has lost much of its energy and high-frequency components to attenuation, indicating many distributed voids and poor-quality masonry.

pulses generated by the mechanical pulse technique were required to successfully penetrate through the walls of the structure.

Figure 8-13 shows a contour plot of pulse "quality" for one of the walls as determined by the equipment operator. A gridwork of direct transmission tests were conducted with 8 inches between centers and the resultant wave forms interpreted qualitatively for amplitude and frequency content. Level 1 readings indicate a large amplitude wave with many of the higher-frequency components intact and correlates to dense unflawed masonry. Level 5 readings, on the other hand, indicate a waveform that has lost much of its energy and high-frequency components to attenuation due to the presence of many distributed cracks and voids, indicating poor quality masonry. The resultant contour plot of Figure 8-13 shows variations in material quality that were not clearly evident by using wave velocity alone. A decline in pulse quality from the top of the wall toward the base indicates internal wall damage increases toward the base of the wall.

8.5.4 *Monitoring Grout Injection*

Atkinson [12] and Berra [9] used ultrasonic testing to monitor grout penetration during projects for repair of damaged masonry. Ultrasonic pulse velocity could be correlated to masonry damage and also was effective at determining the extent of grout flow into voids within the masonry. Berra found sonic tests to be more useful for characterization of severely damaged masonry and developed a guide for predicting masonry repair quality. The received waveform is analyzed in the frequency domain and the product of the amplitude and frequency of the fundamental wave is used as the basis for characterization.

A laboratory test wall was constructed by Atkinson [12] to represent typical old multi-wythe masonry construction. The wall was damaged by compressive and shear overloads and repaired by injection of grout into interior voids and cracks. Through-wall ultrasonic pulse testing was conducted in each of the original, damaged, and repaired states. Results in the form of three-dimensional contour plots of pulse arrival time are shown in Figure 8-14. These plots show that ultrasonic pulse testing was effective in locating damage in the wall interior. Post-repair testing showed that grout injection was quite successful for repairing cracks and voids and restoring the wall to its original structural condition.

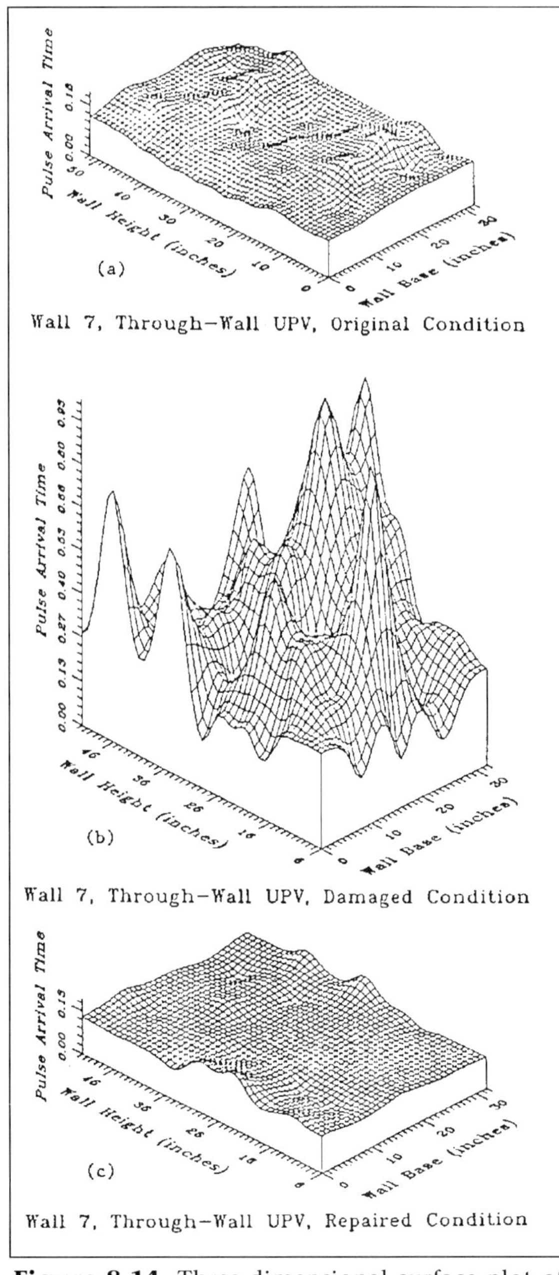

Figure 8-14. Three-dimensional surface plot of through-wall ultrasonic pulse arrival time for a masonry wall. (a) Original, as-built condition; (b) damaged condition; (c) following repair by injection of cementitious grout into internal voids and fractures.

Condition	Mean Pulse Arrival Time (microsec.)
Original	148
Damaged	357
Repaired	111

8.6 *SUMMARY*

Large areas of masonry can be evaluated in a rapid manner using pulse transmission techniques. Transient stress waves traveling through masonry are affected by material properties, voids, cracks, and general condition. The parameter of interest in pulse transmission techniques is usually pulse velocity. Wave attenuation characteristics such as energy and frequency content may be studied to provide additional information on masonry condition. Measurement of pulse transmission is most useful for initial condition surveys for indicating suspect areas that require further investigation. Correlations with material properties are questionable and require companion testing for any reliability.

Ultrasonic pulse velocity testing is a rapid technique best suited for use on modern masonry or relatively unflawed structures. The technique is able to locate flaws with sizes of 2 to 3 inches. Ultrasonic testing is most useful for direct or through-wall applications where the path length is less than about 16 inches and is especially applicable toward evaluation of grouted masonry.

Mechanical pulse testing uses low-frequency, high-amplitude sonic pulses. This type of waveform is not as rapidly attenuated by the numerous boundaries and small flaws found in typical masonry construction and is more useful for evaluation of old and massive masonry structures. No field-portable commercial packages are available and hence sonic testing requires an experienced operator and a significant equipment investment.

Mechanical pulse velocity is more useful as a global technique because the generated pulse is able to travel long distances through the masonry. It is best used as an indicator of overall material quality and uniformity and also locates larger cracks and voids. Mechanical pulse testing is limited to location of voids of 8 to 12 inches and larger.

8.7 REFERENCES

1. Noland, J.L., R.H. Atkinson, and J.C. Baur. 1982. An Investigation into Methods of Nondestructive Evaluation of Masonry Structures. Report to the National Science Foundation, National Technical Information Service Report No. PB 82218074.
2. Hobbs, B., and S.J. Wright. 1987. Ultrasonic Testing for Fault Detection in Brickwork and Blockwork. Proceedings at the International Conference on Structural Faults and Repair. London.
3. Calvi, G.M. 1988. Correlation Between Ultrasonic and Load Tests on Old Masonry Specimens. Proceedings at the 8th International Brick/Block Masonry Conference. Dublin, Ireland.
4. Noland, J.L., R.H. Atkinson, G.R. Kingsley, and M.P. Schuller. 1990. Nondestructive Evaluation of Masonry Structures. Report presented to the National Science Foundation. Atkinson-Noland & Associates Inc. Boulder, Colo.
5. Komeyli-Birjandi, F., M.C. Forde, and H.W. Whittington. 1989. Sonic Investigation of Shear Failed Reinforced Brick Masonry. *Masonry Industry.* November.
6. Forde, M.C. 1992. Nondestructive Evaluation of Masonry Bridges. Proceedings at the Conference on Nondestructive Evaluation of Civil Structures and Materials. Atkinson-Noland & Associates Inc. Boulder, Colo.

7. Carino, N.J. 1984. Laboratory Study of Flaw Detection in Concrete by The Pulse Echo Method, SP82-28. *In-Situ Nondestructive Testing of Concrete.* Detroit: American Concrete Institute.
8. Aerojet General Corporation. 1967. Investigation of Sonic Testing of Masonry Walls. Final report to the Department of General Services, Office of Architecture and Construction, State of California.
9. Berra, M., L. Binda, L. Anti, and A. Fatticioni. 1992. Utilization of Sonic Tests to Evaluate Damaged and Repaired Masonry. Proceedings at the Conference on Nondestructive Evaluation of Civil Structures and Materials. Atkinson-Noland & Associates Inc. Boulder, Colo.
10. Hobbs, B. 1991. Development of a Construction Quality Control Procedure Using Ultrasonic Testing. Proceedings at the 9th International Brick/Block Masonry Conference. Berlin, Germany.
11. Senbu, O., A. Baba, M. Abe, and M. Sugiyama. 1991. Effect of Admixtures on Compatability and Properties of Grout. Proceedings at the 9th International Brick/Block Masonry Conference. Berlin, Germany.
12. Atkinson, R.H., and M.P. Schuller. Evaluation of Injectable Cementitious Grouts for Repair and Retrofit of Masonry. *Masonry: Design and Construction, Problems and Repair,* ASTM STP 1180. Philadelphia: ASTM.
13. ASTM. 1992. *Standard Test Method for Pulse Velocity Through Concrete,* ASTM C 597. Philadelphia: ASTM.

Chapter Nine • Flatjack Tests for In-place Masonry Evaluation

Engineers faced with the challenge of repair or retrofit of existing masonry buildings require knowledge of the loads applied to the masonry, its compressive strength, and compressive deformability properties. Historically, information on masonry compressive behavior has been obtained by destructive removal of test prisms to the laboratory for evaluation. The flatjack method has been developed as an in-place technique to minimize masonry damage during testing. In-place testing using flatjacks allows direct measurement of the state of compressive stress present within the masonry in addition to compressive behavior with a minimum of disruption to the masonry. Flatjack testing requires removal of a portion of a mortar bed joint however can be considered nondestructive because this damage is temporary and is easily repaired after testing.

9.1 BACKGROUND

Flatjack testing was first used in the field of rock mechanics for determining stress states in mines and tunnel linings. Italian researcher Rossi [1] adapted the method for use with masonry in the early 1980s, and since that time the method has been studied in the United States [2], China [3,4], and Europe [1,5,6,7] for application to masonry evaluations. Flatjack testing has been proven for use on solid unreinforced brick and stone masonry. It may be possible to test hollow, grouted, and reinforced masonry using flatjack methods, but these applications have not yet been fully established.

Two separate test standards for masonry evaluation using flatjacks were developed by ASTM and approved in 1991. ASTM Standard Test Methods C 1196-91, *In Situ Compressive Stress Within Solid Unit Masonry Estimated Using Flatjack Measurements,* and C 1197-91, *In Situ Measurement of Masonry Deformability Properties Using the Flatjack Method,* describe the two techniques for masonry evaluation. European practice follows RILEM standards LUM.D.2 and LUM.D.3, which were first introduced in 1990.

Two different types of information can be obtained using the flatjack method: in-place state of compressive stress and masonry compressive behavior.

9.1.1 *In-place Stress Test*

The existing state of compressive stress is measured using the in-place stress test, or single flatjack test. The test is based on a simple principle of stress relief and measures the stress in the masonry from dead loads, any acting live loads, thermally induced stresses, or load transfer resulting from shrinkage of building frames. The advantage of in-place measurement of compressive stress is that no assumptions regarding existing loads, load paths, and the contribution of other structural members are needed to determine the stress in the masonry. Information on the state of vertical stress also can be used in the development of analytical models for structural evaluations. This information is valuable for analysis of arches, vaults, and other complex masonry structures. Stress tests also can be conducted on opposite sides of an eccentrically loaded pier or wall to determine the magnitude of induced bending moments.

The in-place stress test does have limitations: the state of compressive stress is measured only in the wythe being investigated and stress in other wythes of masonry may be different. The principle of the test is based on the assumption that the surrounding masonry is homogeneous and deforms symmetrically around the slot. These assumptions are not strictly correct for masonry, but experimental and analytical investigations [8,9] have shown that the effect of load redistributions and non-proportional deformations are within the accuracy of the method and can be neglected. Laboratory testing has shown that the in-place stress test has a margin of error of up to 20%.

Other methods for in-place determination of compressive stress using strain relief tests have been used [10]. Strain relief tests are conducted by measuring changes in strain on the masonry surface induced by coring or removal of a mortar joint. The main disadvantage of this method is that masonry elastic properties must be assumed or measured using other methods to calculate in-place stress states.

9.1.2 *In-place Deformability Test*

The in-place deformability test uses two parallel flatjacks, separated vertically by several courses of masonry, to subject the masonry to compressive stresses. Corresponding strains are measured using surface-mounted compressometers to provide a measure of the compressive behavior of the masonry. The compressive modulus is calculated from the experimental stress-strain curve and the masonry can be loaded to failure to determine compressive strength if this type of damage is acceptable. It is not always desirable to fail the masonry for obvious aesthetic and structural considerations, and loading is usually terminated when the stress-strain curve begins to level out, indicating impending failure. The peak compressive strength then can be estimated based on the shape of the stress-strain curve.

Compressive behavior is measured only on the wythe of masonry being investigated and the deformability of other wythes may be different. The masonry between flatjacks is not subjected to a uniform compressive stress: boundary effects due to the collar joint and surrounding masonry partially restrain the loaded masonry, but these effects are negligible toward the middle of the loaded area. Experimental verification tests and analytical models have shown the method generally overestimates the masonry compressive modulus by up to 15% to 20% [9,11]. Comparisons between tests conducted using flatjacks and tests on prisms removed from masonry [12, 20] shows that the flatjack test provides a reasonable measure of masonry compressive behavior.

9.2 EQUIPMENT

Equipment for flatjack testing is neither prohibitively expensive nor does it require operation by highly experienced personnel. The basic equipment consists of:

- Mortar removal equipment (drill or masonry saw)
- Two or more flatjacks
- Flatjack shims
- Hydraulic pump, pressure gage, and hoses
- Dial gages or electronic LVDTs for deformation measurement
- Safety equipment

9.2.1 Mortar Removal

A portion of a masonry bed joint must be cleared to provide a space for flatjack insertion. This is best accomplished using a masonry saw with a water-cooled diamond blade. On older weak mortars it may be easier to remove the mortar by stitch drilling, that is, successive drilling of closely spaced holes into the joint. A carbide tipped or hollow-core masonry drill bit is recommended for these cases. High-power hammer drills or power chisels should be avoided because the vibration from such equipment may weaken the bond of adjacent mortar joints.

9.2.2 Flatjacks

A flatjack is a thin steel envelope sealed at the edges and fitted with ports, allowing internal pressurization using hydraulic equipment. Flatjacks can be manufactured in any size or configuration, but several specific designs are normally used for masonry evaluations. The flatjacks shown in Figure 9-1 are all manufactured with a thickness no greater than 0.375 inch to fit within a mortar joint and have a width equal to the width of masonry unit being loaded. The rectangular flatjacks are used in slots where mortar was removed by stitch drilling, and the semi-circular jacks have a radius on each end to fit snugly in a slot prepared by cutting with a masonry saw. Rectangular and extended semi-circular jacks having a length equal to two or more masonry units (Figure 9-1 (a) and (b)) can be used for either the in-place stress test or the in-place deformability test. The shorter rectangular and semi-circular jacks (Figure 9-1 (c) and (d)) are useful for only the in-place stress test.

Flatjacks are manufactured to be flexible and provide a linear output of stress over working pressure range. Flatjacks manufactured for masonry testing have a maximum pressure rating of 1000 psi. Internal stiffness of the jack and flexibility of edge welds necessitates calibration of flatjacks before use to determine the relationship between internal pressure and stress applied to the masonry. The calibration procedure involves measurement of the stress output of the flatjack as a function of internal

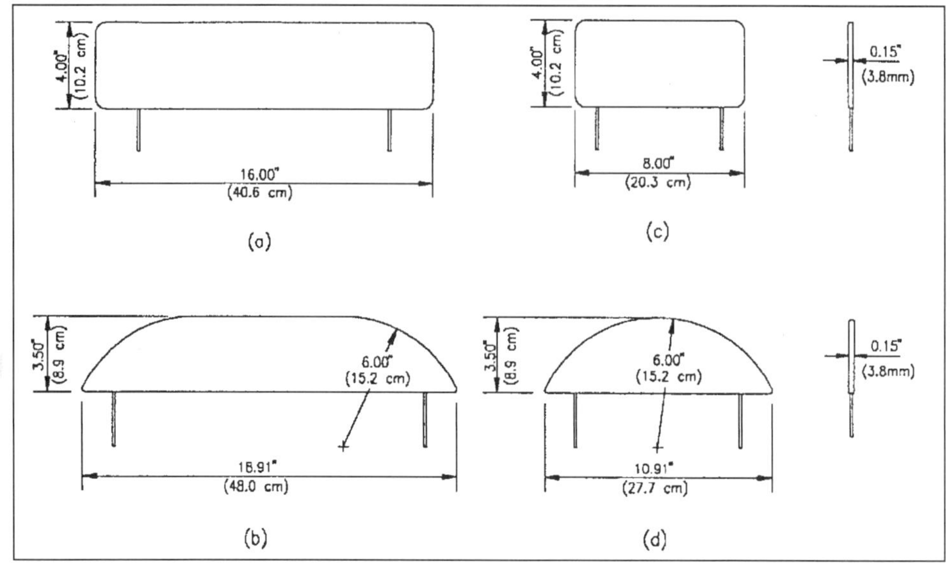

Figure 9-1 Different flatjack configurations. Types (a) and (c) are used where masonry was removed by routing or stitch drilling; types (b) and (d) are used in slots formed by cutting the mortar with a circular saw.

hydraulic pressure. The calibration constant K_m will be less than 1.0 and is used as a modifier to the measured hydraulic pressure. Both ASTM C 1196 and C 1197 contain a complete description of the calibration procedure. Flatjacks soften with repeated usage and should be re-calibrated after about five tests or if the jack has undergone extreme deformations.

9.2.3 *Shims*

The test flatjack must be kept at near its original thickness during pressurization to prevent the calibration factor from changing. Hence, before pressurization begins, shims must be placed around the test flatjack to completely fill the slot. Shims also are used to bridge any small gaps or voids that may damage the test flatjack and aid in removal of the test jack after testing. Grouting of the flatjack in the slot is not allowed as the grout may flow into the collar joint or adjacent voids, altering masonry compressive behavior.

Shims should have the same shape and size as the flatjack being used. Stiff metal shims may be used if the slot has a constant thickness; use multipiece shims for irregular slots. Use of stiff metal shims is not preferred because this practice can cause an erroneous stiffening behavior under compressive stresses as the shims deform under low stresses. The most uniform transfer of stress from the test jack to the surrounding masonry can be obtained when additional flatjacks are used as "fluid cushion" shims as shown in Figure 9-2. One or more additional flatjacks are placed around the test jack and seated by pressurizing to about 80%

Figure 9-2. Additional flatjacks are placed around the working jack to fill the slot and act as fluid cushion shims. Fluid shims deform into surface irregularities and provide the most consistent transfer of stress from the working jack to the masonry.

of the estimated maximum load. This forces the jack to deform into small irregularities in the slot and helps ensure full contact with the surrounding masonry. The use of shims is discussed in an annex to ASTM C 1196-91 and C 1197-91 on masonry testing.

9.2.4 *Hydraulic Equipment*

A hydraulic pump is used to pressurize flatjacks. Either manual or electronically actuated pumps are adequate. A test-quality pressure gage with a 4- or 6-inch-diameter dial and a minimum accuracy of 1% of full scale should be used for testing applications. The gage should have a range approximately equal to the flatjack capacity.

Hydraulic hoses are used to connect the pump to the jacks. To avoid oil spills during connection of hydraulic hoses, use flatjacks fitted with valves or quick-connect fittings. Hydraulic equipment using water or other non-staining fluid is recommended for use when testing structures with sensitive architectural finishes.

9.2.5 *Deformation Measurement*

The in-place stress test requires deformation measuring instruments that are not rigidly fixed to the masonry surface: readings are taken before and after the slot is prepared, hence any instrumentation cannot interfere with mortar removal equipment. A multilength removable extensometer such as a Whittemore gage is most suitable for these measurements. The measurement device should have a minimum resolution of 0.0002 inch.

Instrumentation may be fixed to the surface of the masonry after preparing slots for the in-place deformability test. The removable Whittemore gage may be used during deformability testing, however the large number of readings required makes this task quite tedious. More useful are surface mounted gages. Where electricity is not readily available, dial gages provide an inexpensive alternative. A dial gage with an accuracy of 0.0002 inch is sufficient for conducting deformability tests of masonry.

Electronic devices such as linear variable differential transformers (LVDT's) can be mounted on the masonry surface and provide a resolution of better than 0.0001 inch. LVDT's usually cost about twice as much as dial gages with similar accuracy, but the use of such devices greatly decrease the actual test time and do not require rigid monitoring by the test operator. LVDT's require either an AC or DC power source and some method for measuring voltage output. Voltage output may be read manually using a digital multimeter; however, the preferable technique is connection to a data acquisition device and portable computer. A system set up in this fashion acquires data automatically during the test, allowing real-time display of a load-displacement or stress-strain curve. The formatted data can be written directly to the computer for efficient post-test plotting and data reduction.

9.2.6 *Safety Equipment*

As with any high-pressure operation, necessary protection against failure of hydraulic equipment must be provided. Gloves and eye protection or a face mask are recommended to protect the operator in the case of equipment failure. During slot preparation, gloves, and eye and face protection are required; a dust mask or respirator also may be necessary.

9.3 *TEST PROCEDURE*

Test preparations for both the in-place stress test and in-place deformability test are similar. The basic test setup is described first, followed by specific procedures for each test.

9.3.1 *Test Locations*

Locations for flatjack testing should be chosen by the engineer or architect based on desired objectives. Stress tests conducted in the vicinity of wall openings, changes in cross-section, or other stress concentrations may not provide consistent results and should be avoided. In-place deformability tests must be conducted at locations where a sufficient mass exists above the flatjacks to resist applied stresses. Large stresses applied during deformability testing often cause noticeable uplift cracking next to flatjacks in low-rise or lightly stressed masonry structures.

Specific locations for flatjack testing should be chosen to obtain the most consistent transfer of stress from the jack to the masonry: avoid areas with irregular bed joints, excessive voids, or broken units. If possible, do not include header courses within the test area.

It is not necessary to load the entire wall thickness for either type of flatjack test. More reliable results are obtained if the entire wall thickness is loaded, but this is not always possible when testing massive multi-wythe or stone masonry structures. Flatjacks should be sized to load a minimum of one wythe of masonry in any case.

Multiwythe masonry walls have a variation in material properties and construction quality throughout the wall width. The exposed wythe on the building's exterior is usually constructed with a high-quality face brick and consistent quality to minimize variations in architectural appearance. Masonry on the wall's interior or masonry covered by surface treatments is often of an entirely different composition and may consist of lower-quality common brick, stone, or rubble construction. The interior wythes are structurally significant and in massive bearing wall structures may comprise the majority of the structural element. Because of these differences it is advisable to conduct deformability tests on both exterior an interior wythes. This entails removing plaster or a portion of the exterior wythe to gain access to the wall interior.

Flatjacks for masonry testing are manufactured to be thin enough to fit within a masonry bed joint, hence most tests are conducted to subject the masonry to vertical stresses. If required by the engineer, information on horizontal or diagonal compressive behavior also may be obtained using flatjacks. Tests can be conducted with the flatjacks in a vertical or diagonal configuration if saw cutting of the units is allowable and this information is of use for analysis.

Due to the inherent variability of results obtained with flatjack testing, conduct at least three to five tests in each general area of interest to obtain a statistical sample.

9.3.2 *Slot Preparation*

Once the test area has been chosen, carefully mark the slot, indicating the mortar to be removed. All mortar must be removed from the joint to provide a smooth bearing surface for the flatjack. This is most easily accomplished by using a water-cooled masonry saw with a diamond-impregnated blade. Weak mortars may be removed best by stitch drilling, followed by removal of any remaining mortar using a chisel and hammer. Use an air hose or vacuum to remove all particles from the slot before flatjack insertion.

9.3.3 *Flatjacks and Shims*

It is essential that the test jack be kept near its original calibrated thickness during the test. Shims, described in the equipment section of this chapter, are used to fill the slot around the loading flatjack and to bridge small internal voids. After the flatjacks and shims are placed in the slot, the flatjack is "seated" by pressurizing to a minimum of 200 psi or 80% of the expected maximum pressure using three separate cycles. The seating process forces the flatjack to conform to any small irregularities in the slot to minimize stress concentrations and flatjack deformations during the actual test.

9.4 *IN-PLACE STRESS TEST*

9.4.1 *Initial Measurements*

The technique for the in-place stress test requires initial displacement measurements to be made before removing mortar from the flatjack slot. In general, three to ten separate lines of gage points are used. One possible test setup, showing seven lines of measurement distributed over the slot, is shown in Figure 9-3. It is best to avoid placing gage points on the units next to the slot as these units may be disturbed slightly during mortar removal. Use epoxy to fasten the gage points firmly to the masonry and avoid placing gage points in mortar joints.

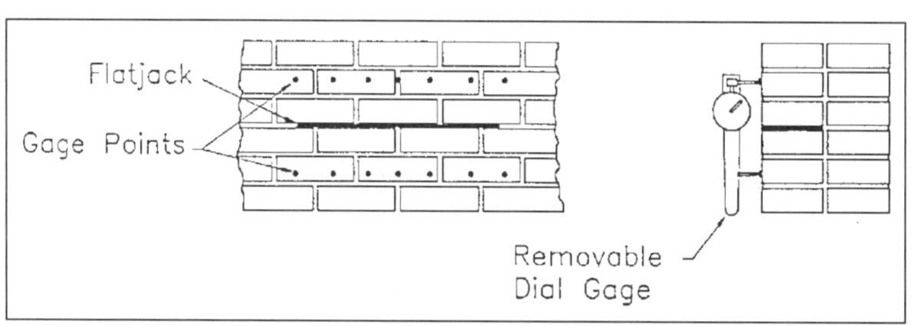

Figure 9-3. Test setup for in-place determination of the state of compressive stress present within the masonry.

Pre-test measurements are used to determine the initial location of the gage points. Several repetitions of the initial reading should be made as this information is important and forms the baseline for future comparisons. Mortar is removed as described above following completion of the initial measurements and it is very important not to disturb the gage points or units themselves during this process.

9.4.2 *Stress Restoration*

Following mortar removal from the flatjack slot, another series of measurements are made. The separation between gage points will have decreased due to stress relief caused by removal of bed joint mortar. Flatjacks are inserted and seated as described above; pressure is increased in small increments of 10 to 20 psi while gage point separations are monitored. Most old load-bearing masonry has a compressive stress ranging from less than 10 psi up to 100 psi. In-place stresses of 150 psi and more have been recorded for over-stressed masonry. Compressive stresses in modern reinforced masonry can be as high as 300 to 500 psi.

When the original separation between gage points has been restored, the internal hydraulic pressure of the flatjack is recorded. This pressure is modified by the calibration constant and an area constant to indicate the

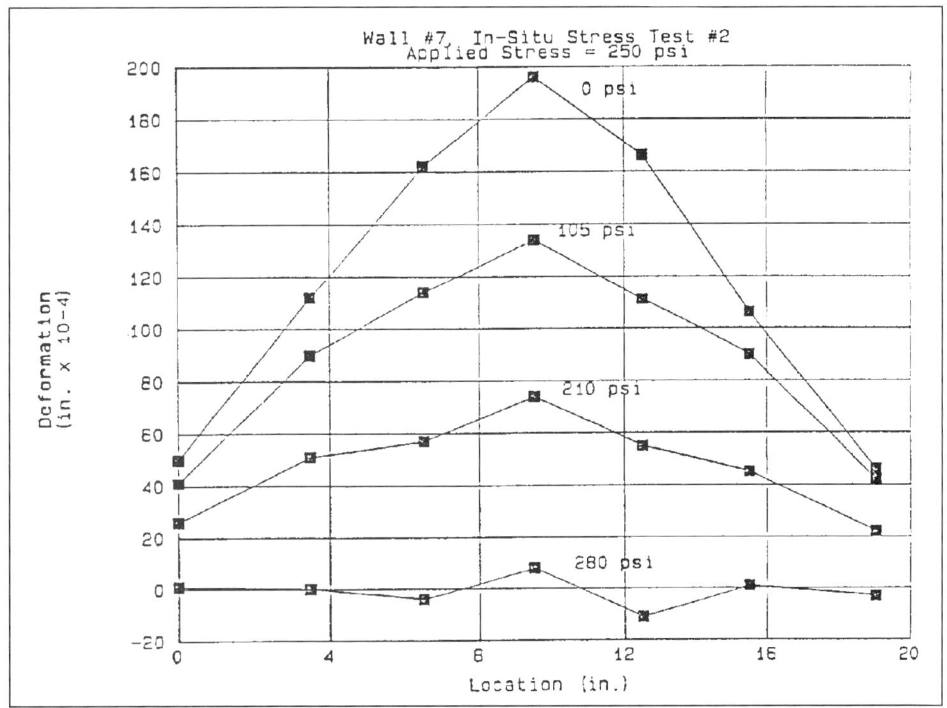

Figure 9-4. Typical test data from the in-place stress test showing masonry deformations around the slot for various levels of flatjack pressure. The cancellation stress of 280 psi indicates the compressive stress in the masonry at this location.

state of compressive stress present in the masonry at the test location (see section on analysis). A typical series of measurements are shown in Figure 9-4, where the individual curves show the variation in gage point separation for the indicated stress levels. In this case an average compressive stress of 280 psi has been measured.

When the restoration stress has been determined, flatjack pressure is reduced to zero. The flatjack should remain in place if deformability tests are to be conducted, in which case a second flatjack is installed several courses away from the original flatjack. Proceeding in this manner makes the most efficient usage of labor required for removing mortar and setting up the test. Between four and eight man hours are required for each in-place stress test, depending largely on the condition of the mortar and time required for slot preparation. The actual test can be conducted in less than one hour.

9.5 *IN-PLACE DEFORMABILITY TEST*

9.5.1 *Deformation Measurement*

Two parallel flatjacks are used for the in-place deformability test. The separation between jacks is no greater than the length of the jack and nor-

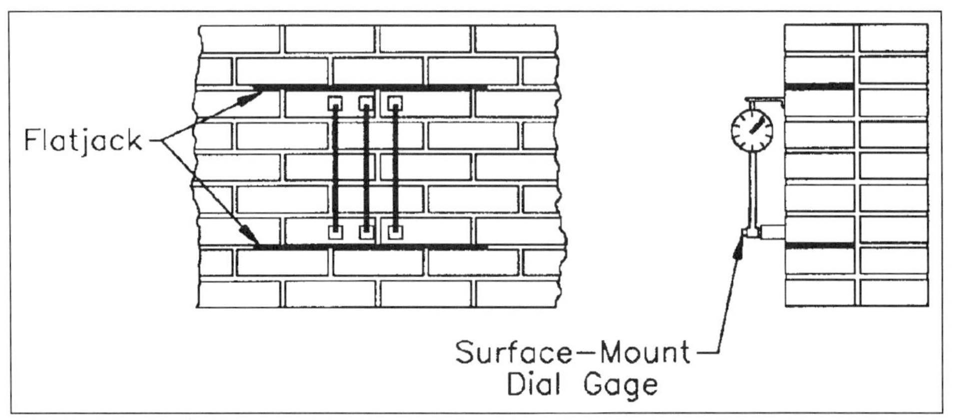

Figure 9-5. Equipment setup for the in-place deformability test.

mally is about three to five courses of masonry. Several vertical lines of deformation measurements are needed, as shown in Figure 9-5. Analytical investigations [9] have shown that the state of compressive stress is most consistent within the middle half of the flatjack length; outside this area the stress state is affected by restraint from adjacent masonry and may lead to erroneous deformation readings. A series of three measurement devices located within the middle third of the flatjack length are used; data from the three gages are averaged to obtain strain information.

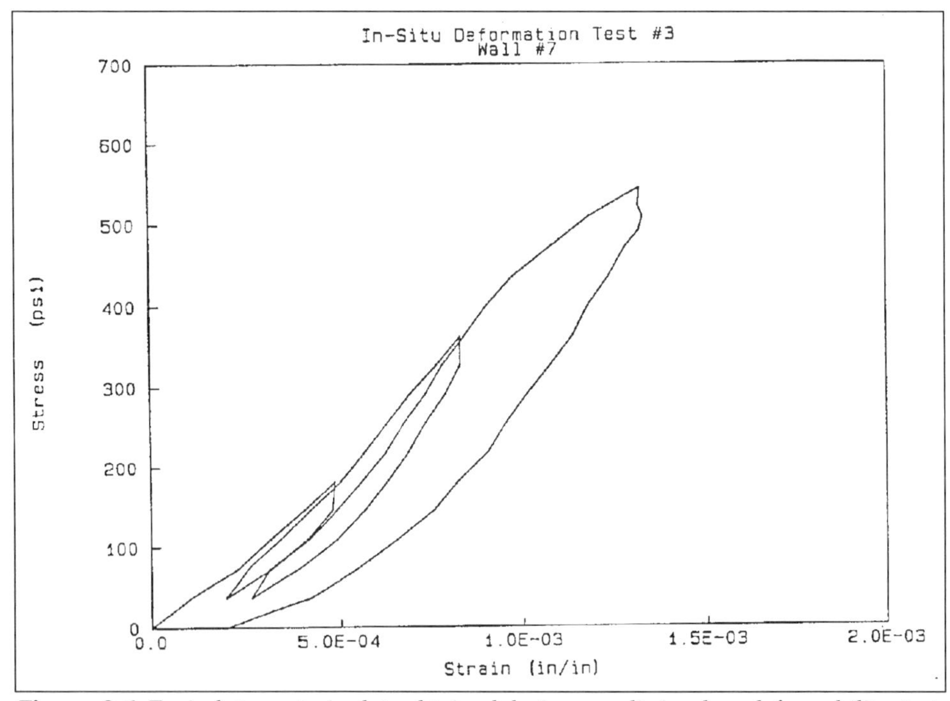

Figure 9-6. Typical stress-strain data obtained during a cyclic in-place deformability test.

9.5.2 *Stress Application*

The flatjacks are pressurized in increments of 5 to 20 psi. Deformation measurements are taken and recorded for each stress level. A plot of the stress-strain or load-displacement behavior should be maintained during loading. If damage to the masonry is acceptable, the test may be carried to failure. It is usually preferable to monitor the stress-strain behavior during loading and terminate the test when the curve begins to become highly non-linear, indicating imminent failure. An indication of cyclic behavior may be obtained by a series of repeated loadings, as shown in Figure 9-6.

Most older masonry has a peak compressive strength ranging from 300 psi to 1500 psi and more. Values for the initial compressive modulus typically range from 70 ksi to 1500 ksi, depending on unit type, mortar quality, and general masonry deterioration.

9.5.3 *Post-test Repairs*

Following either of the flatjack tests, the pressure in the jacks is reduced to zero and the jacks removed from the slot. The slot is flushed with water and a mortar similar in color and composition to the original mortar is pointed into the slot, restoring the masonry to its original condition.

9.5.4 *Labor Requirements*

The major portion of labor associated with the in-place deformability test is involved with slot preparation and equipment setup. Comparable to the

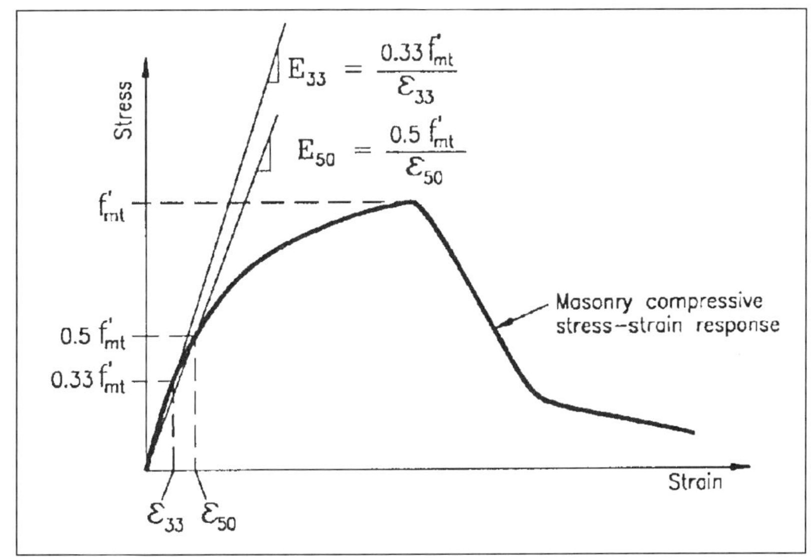

Figure 9-7. Compressive modulus is defined as the slope of the initial portion of the stress-strain curve and may be determined using linear regression or a chord modulus from 5% to 33% or 50% of the compressive strength.

in-place stress test, this preparation requires between four and six man hours. Load application and deformation measurement requires less than one hour. Post-test repairs need less than one-half hour for most situations.

9.6 ANALYSIS/INTERPRETATION OF RESULTS

9.6.1 Converting Flatjack Pressure to Stress

Internal flatjack pressure must be modified by two factors to determine the actual stress applied to the masonry. The calibration factor K_m relates internal flatjack pressure to average stress output of the jack and is normally supplied by the manufacturer. The area factor K_a is the ratio of the flatjack area divided by the slot area and must be determined for each test location. The actual stress applied to the masonry, σ, is calculated by multiplying the measured hydraulic pressure P by these two constants:

$$\sigma = P^* K_m K_a \quad (9\text{-}1)$$

9.6.2 Compressive Modulus

The main goal of the in-place deformability test is to obtain a measure of the compressive modulus of the masonry. Several different measures of modulus may be used, as shown in Figure 9-7, and the measure used should be identified in the test report. An initial stiffening portion may

Figure 9-8a. The Seney-Stovall Chapel, constructed in 1892, is located on the University of Georgia campus in Athens, Ga.

indicate closing of small cracks or flatjack deformation at low stresses and is normally disregarded. The initial linear portion of the stress-strain curve may be fitted with a straight line using linear regression to indicate compressive modulus. A value more useful to engineering analyzes the chord modulus to either 33% or 50% of the peak compressive strength. This measure is obtained by calculating the slope of a line drawn between points on the stress-strain curve representing 5% and either 33% or 50% of the compressive strength. The initial point of 5% is used to avoid any initial non-linearity of the curve as described above.

9.6.3 *Estimation of Compressive Strength*

It is possible to load the masonry to failure during the in-place deformability test, but it is usually desirable to terminate loading before the peak compressive strength has been reached to avoid excessive damage to the masonry. Several analytical procedures have been developed to describe the compressive behavior of masonry using mathematical formula [13,14]. These techniques can be used to estimate masonry compressive behavior based on the initial modulus and curvature of the stress-strain curve. These descriptive mathematical models have been developed for specific combinations of unit and mortar types and may not be applicable for all types of masonry. Such techniques should be used with caution and any information from extrapolations of test data should be clearly identified as being an approximation or estimate of peak strength.

Figure 9-8b. The interior view is from the stage; the masonry wall on stage left supports several roof trusses.

A simple technique for estimation of the peak strength is to use commercially available curve-fitting software. A polynomial curve can be fit to test data and extrapolated to determine peak compressive strength. Reliable

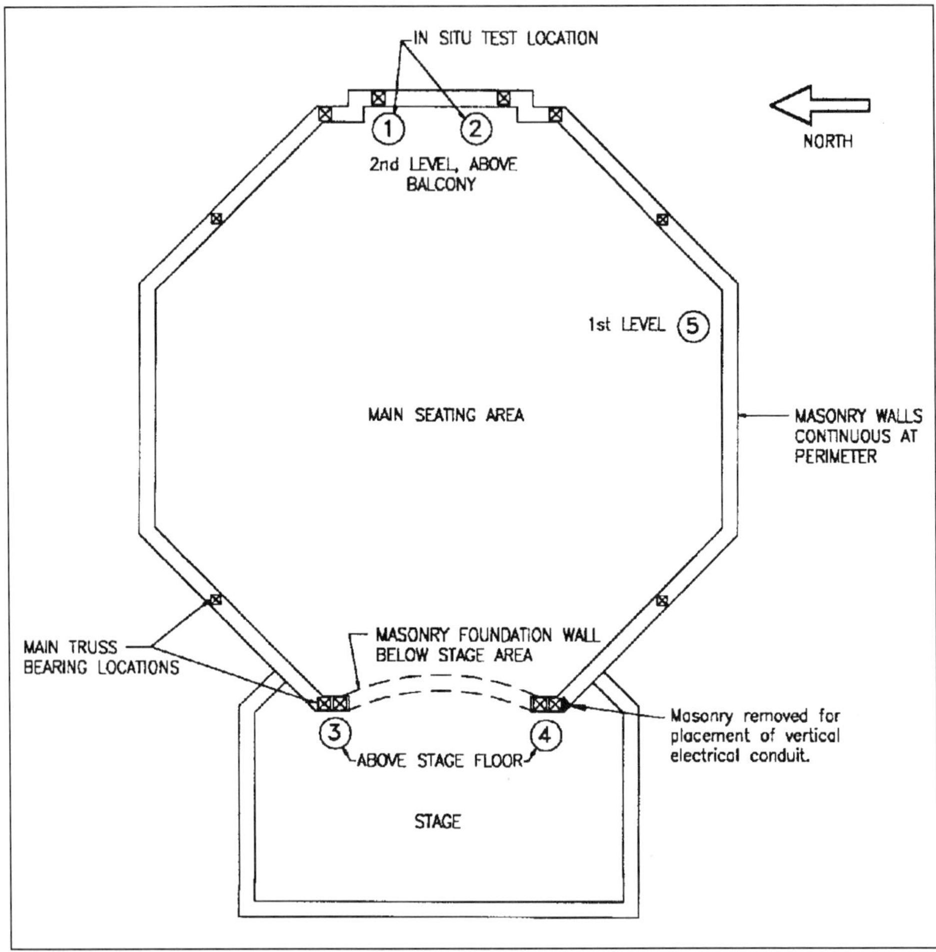

Figure 9-9. Seney-Stovall Chapel — plan view.

estimates of compressive strength can be obtained if the fitted curve conforms reasonably well to the shape of the test data and if the masonry has been loaded well into the inelastic range.

9.7 *APPLICATIONS*

9.7.1 *In-place Stress Variations*

One of the more useful applications of the in-place stress test is as a measure of stress gradients, which indicate bending moments or changing load conditions. Two in-place stress tests were conducted on a chapel constructed in 1892 in Athens, Ga. (Figure 9-8). Roof loads were carried by a series of large wooden trusses to masonry bearing walls around the perimeter of the octagonal chapel. The main stage, situated along one edge of the seating area as shown in Figure 9-9, interrupts the bearing

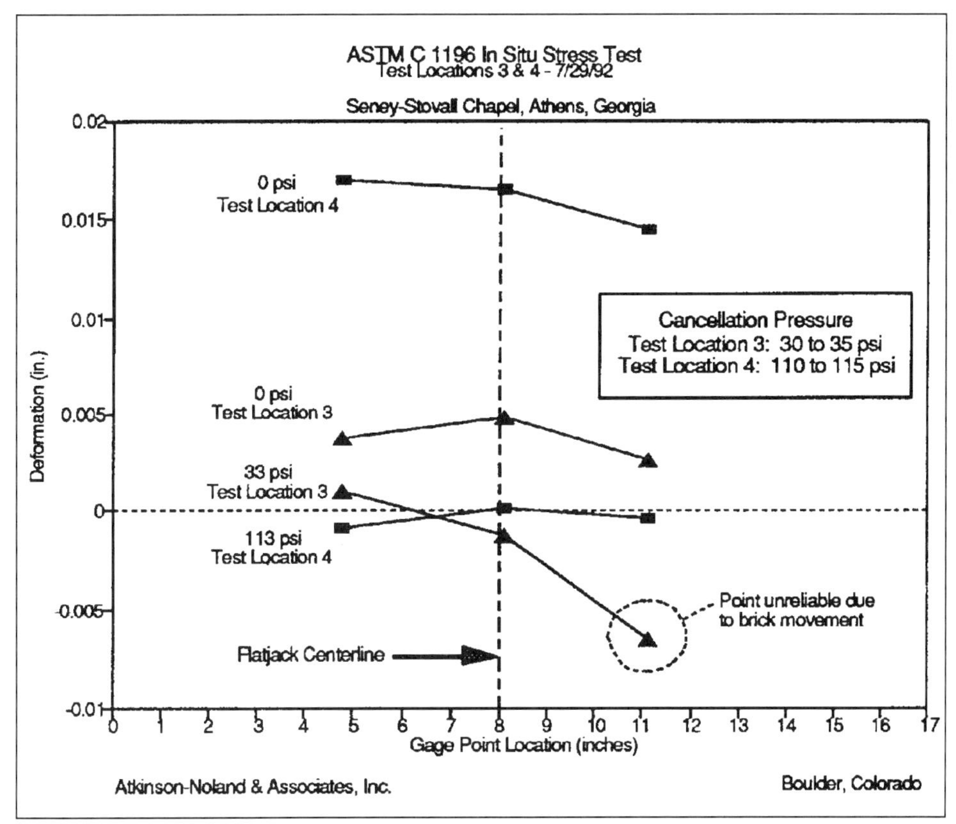

Figure 9-10. Results for the in-place stress test conducted on both sides of the stage in the chapel. A compressive stress of between 30 and 35 psi exists at stage left; a compressive stress of between 110 and 115 psi was measured at stage right.

walls and concentrates the roof loads onto two masonry stub walls on either side of the stage. Results of the in-place stress test, shown in Figure 9-10, indicated that although masonry stress at one side of the stage was within predicted levels, the wall on the other side of the stage was overstressed. Further investigation revealed that the over-stressed stub wall was effectively isolated by the removal of a vertical series of bricks for placement of an electrical conduit, preventing transfer of stresses to adjacent masonry. Hence the investigation indicated the necessity for additional shoring of the roof trusses at this location.

9.7.2 *Eccentric Wall Loadings*

Many buildings experience foundation settlement and shifting of structural elements over time. The in-place stress test was used to measure masonry compressive stresses resulting from such settlement in a 200-year-old historical masonry building in New Orleans. The building was originally founded on wooden pilings resting directly on Mississippi River silt.

Foundation movement and redistribution of stresses resulting from removal of interior shear walls resulted in tilting of main bearing walls, measured to be 22 inches out of plumb in only two stories of height. In-place stress tests were conducted on the wall shown in Figures 9-11a and 9-11b to determine stress eccentricities and bending moments resulting from wall tilting. Tests conducted on the interior (downhill) side of the wall measured a compressive stress of between 70 psi and 100 psi, or about two times the expected dead load stresses. No deformations were measured following the cutting of the flatjack slot on the wall exterior (uphill side), indicating no compressive stress at this surface. Development of tensile stresses on the uphill wall side can be expected with further wall movement.

Figure 9-11a. Several bearing walls of an historic New Orleans building were leaning towards the building interior.

9.7.3 *Compressive Stresses in Masonry Veneer*

Design of modern masonry veneer structures should incorporate movement joints at shelf angle connections to account for thermal movements and shortening of the structural frame over time. Older structures with stone, masonry, or terra-cotta veneer may have been constructed without such provisions. The lack of these joints often leads to the development of significant compressive stresses and resulting damage to masonry veneer or cladding. Strain relief tests conducted on the terra cotta cladding [10] of a 27-story steel-frame structure constructed in 1924 were used for determination of relief joint placements. Compressive stresses of up to 800 psi were measured, and recommendations for relief joint placements were made based on experimental results.

Figure 9-11b. In-place stress testing determined the masonry compressive stress to be between 70 and 100 psi in the interior surface. No compressive stress could be measured at the wall's exterior surface, indicating the possible development of tensile stresses with any further wall movement.

9.7.4 *Compressive Behavior of Face Brick versus Interior Wythes*

In-place deformability tests may be used to indicate variations in construction quality throughout a structure. The Mt. St. Gertrude's Academy building is located in Boulder, Colo., and was constructed in 1892 as a three-story load-bearing stone and brick masonry structure. The building was severely damaged by fire in 1980 (Figure 9-12), and a series of in-place deformability tests were conducted to determine material properties of the remaining brickwork.

The investigation determined there was no significant fire damage to the masonry investigated, but it did indicate variations in masonry

Figure 9-12. Mt. St. Gertrude's Academy, constructed in 1892 in Boulder, Colo.

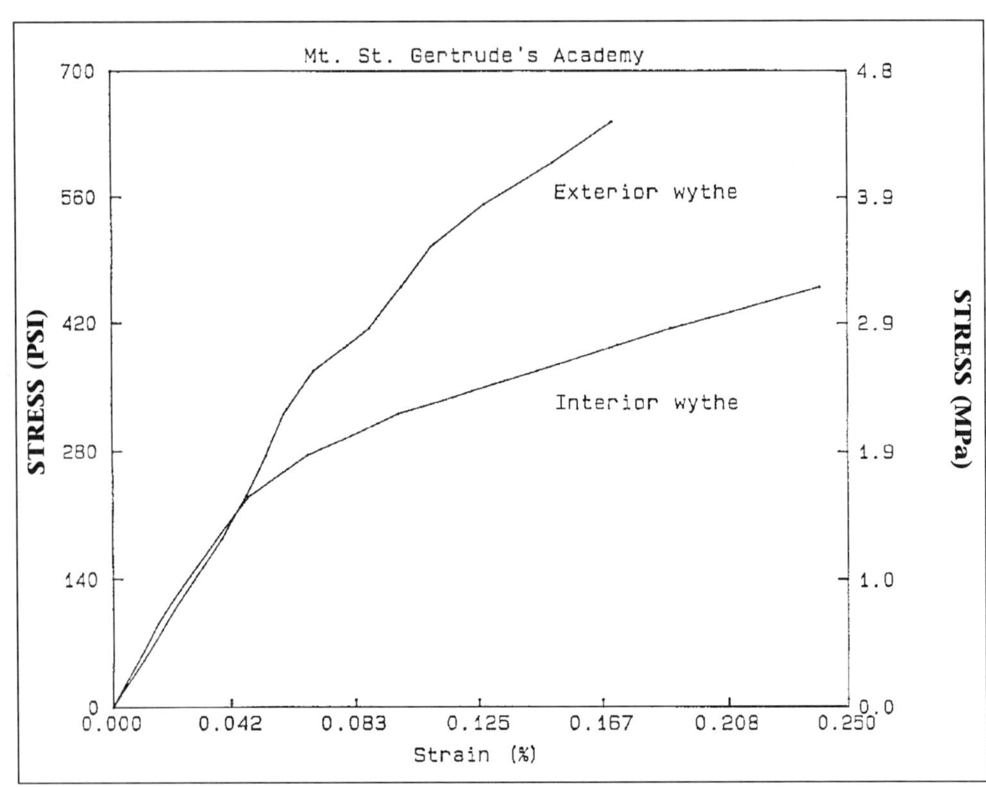

Figure 9-13. In-place deformability test results conducted at Mt. St. Gertrude's Academy show there to be a critical difference in compressive behavior of the face wythe and interior wythes.

quality throughout the structure. The outer wythe of the exterior envelope was constructed with high-quality face brick and superior workmanship. Interior wythes, however, were covered with plaster and hence a lower-quality brickwork was constructed using common units. In-place deformability tests conducted on both the exterior face brick and interior wythes show a marked difference in compressive behavior (Figure 9-13). Variations in material quality as indicated by flatjack testing provide important information for engineering analyses of multiwythe walls.

9.7.5 *Long-term Monitoring*

During renovation of the "Classense" Library in Ravenna, Italy [5] in-place stress tests were conducted on masonry pillars supporting arches. Flatjacks were left in place following completion of the test to monitor changes in stress state. The flatjacks acted as pressure cells and were used to measure changes in the stress state during restoration of foundations and opening of the arches.

9.7.6 *Verification of Repairs*

Rossi [7] conducted flatjack tests on the Wall of Leonardo's "Last Supper" in the Refectory of S. Maria delle Grazie in Milan, Italy. In-place stress tests were conducted on both sides of the wall to check the magnitude of bending moments resulting from load eccentricities. In-place deformability tests were conducted at different locations within the wall to determine the effectiveness of repairs to bomb damage conducted in 1948.

9.8 *SUMMARY*

Flatjack methods provide a reliable technique for in-place determination of existing stresses and masonry compressive behavior. Flatjack tests are simple in principle and require only basic hydraulic and deformation measurement equipment. The two flatjack techniques provide a powerful diagnostic tool for evaluating existing masonry and provide information essential for engineering analyses. Flatjack testing also can be used to indicate variations in material quality throughout a structure and for calibration of nondestructive tests to material compressive properties.

In-place deformability tests measure the compressive stress-strain behavior of masonry. Compressive modulus and, in some cases, compressive strength can be determined by the deformability test. This information is essential for both basic engineering analysis and development of advanced analytical models.

In-place stress tests are used for determination of masonry stresses resulting from dead loads, existing live loads, and bending moments. Information from the in-place stress test can be used to verify or calibrate analytical models. The level of stress present in the masonry is used for determination of masonry stability and safety against collapse.

Flatjack tests are considered nondestructive or semidestructive because a portion of a mortar bed joint must be removed for flatjack insertion. The mortar can be easily replaced following completion of testing to restore the visual appearance of the masonry.

9.9 **REFERENCES**

1. Rossi, P.P. 1982. Analysis of Mechanical Characteristics of Brick Masonry Tested by Means of Nondestructive In Situ Tests. Presented at 6th International Brick Masonry Conference. Rome, Italy.
2. Atkinson, R.H., J.L. Noland, and M.P. Schuller. 1990. A Review of the Flatjack Method for Nondestructive Evaluation. Proceedings at the Conference on Nondestructive Evaluation of Civil Structures and Materials. Atkinson-Noland & Associates Inc. Boulder, Colo.
3. Shi, C., and X. Wang. Analysis and Determination of Compressive Strength of Brickwork in Brick Masonry Walls. Proceedings at the 8th International Brick/Block Masonry Conference. Dublin, Ireland.
4. Wang, Q., and X. Wang. 1988. The Evaluation of Compressive Strength of Brick Masonry In Situ. Proceedings at the 8th International Brick/Block Masonry

Conference. Dublin, Ireland.

5. Binda Maier, L., P.P. Rossi, and G. Sacchi Landriani. 1983. Diagnostic Analysis of Masonry Buildings. International Association for Bridge and Structural Engineering Symposium on Strengthening of Building Structures — Diagnosis and Therapy. Venice, Italy.
6. Rossi, P.P. 1987. Recent Developments of the Flatjack Test on Masonry Structures. Joint USA-Italy Workshop, Evaluation and Retrofit of Masonry Structures. The Masonry Society.
7. Rossi, P.P. 1990. Nondestructive Evaluation of the Mechanical Characteristics of Masonry Structures. Proceedings at the Conference on Nondestructive Evaluation of Civil Structures and Materials. Atkinson-Noland & Associates Inc. Boulder, Colorado.
8. Abdunur, C. 1983. Stress and Deformability in Concrete and Masonry. International Association for Bridge and Structural Engineering Symposium on Strengthening of Building Structures — Diagnosis and Therapy. Venice, Italy.
9. Sacchi Landriani, G., and A. Taliercio. 1986. Numerical Analysis of the Flatjack Test on Masonry Walls. *Journal of Theoretical and Applied Mechanics.* Vol. 5, No. 3.
10. Manmohan, D., R. Schwein, and L. Wyllie. 1988. In Situ Evaluation of Compressive Stresses. *Masonry: Materials, Design, Construction, and Maintenance,* ASTM STP 992. Philadelphia: ASTM.
11. Noland, J.L., R.H. Atkinson, G.R. Kingsley, and M.P. Schuller. 1990. Nondestructive Evaluation of Masonry. Report to the National Science Foundation. Atkinson-Noland & Associates Inc. Boulder, Colo.
12. Turner, S. 1992. Evaluation of the Flatjack Test and In Situ Shear Test of Masonry. Master of Science Thesis. Georgia Institute of Technology. Atlanta.
13. Naraine, K., and S. Sinha. 1989. Loading and Unloading Stress-Strain Curves for Brick Masonry. *Journal of Structural Engineering.* Vol. 115, No. 10. New York: American Society of Civil Engineers.
14. Binda, L., A. Fontana, and G. Frigerio. 1988. Mechanical Behavior of Brick Masonries Derived from Unit and Mortar Characteristics. Proceedings at the 8th International Brick/Block Masonry Conference. Dublin, Ireland.
15. ASTM. 1991. *Standard Test Method for In Situ Compressive Stress Within Solid Unit Masonry Estimated Using Flatjack Measurements,* ASTM C 1196-91. Philadelphia: ASTM.
16. ASTM. 1991. *Standard Test Method for In Situ Measurement of Masonry Deformability Properties Using the Flatjack Method,* ASTM C 1197-91. Philadelphia: ASTM.
17. RILEM. 1990. *LUM.D.2, In Situ Stress Tests Based on the Flatjack.* Paris: The International Union of Testing and Research Laboratories for Materials and Structures.
18. RILEM. 1990. *LUM.D.3, In Situ Strength and Elasticity Tests Based on the Flatjack.* Paris: The International Union of Testing and Research Laboratories for Materials and Structures.
19. Hughes, T., and R. Pritchard. 1994. An Investigation of the Significance of Flatjack Flexibility in the Determination of In Situ Stresses. Proceedings at the 10th International Brick and Block Masonry Conference. Calgary, Alberta, Canada.
20. Kahn, L., and S. Turner. 1985. Evaluation of the Flatjack Test of Brick Masonry. *The Masonry Society Journal.* February.

Chapter Ten • *Analyzing Wet Walls*

A wall is wet and it isn't known why. To find the cause and location of the leak, the wall probably will have to be tested. Although most tests are performed by testing agencies, contractors should know what types of nondestructive tests are available and how those tests work.

10.1 *THE PLASTIC TEST*

Figure 10-1. The tube test can be used to evaluate water permeance. The slower the water in the tube penetrates the wall, the less permeable the wall.

Taping plastic sheeting to the exterior wall can rule out exterior rain penetration. Any moisture that collects on the inside of the sheeting most likely is condensation.

If moisture isn't found, however, condensation can't be ruled out because it's possible for air to leak through the plastic sheet and cause condensation. However the plastic sheet stops almost all air leakage, reducing the potential for condensation. The test usually takes 24 hours.

10.2 *THE TUBE TEST*

Developed by the European organization RILEM, the tube test [1] evaluates the wall's permeability. Reusable tubes, imported from Europe, are available for vertical surfaces and horizontal surfaces.

For vertical surfaces, the tube has a flat circular brim at the bottom that is attached to the substrate with putty. Water is then added to the graduated tube (Figure 10-1). The water level is checked at 5, 10, 15, 20, 30, and 60 minutes and the readings are plotted on a graph comparing the volume of water absorbed with the time required to absorb it. The slower the water penetration, the less permeable the wall.

This test also can be used to test the effectiveness of water repellents. The wall is tested before and after the water repellent is applied and the readings are compared. The test can be completed in about an hour, and eight tests can be performed with one tube in one day.

10.3 *THE SPRAY TEST*

A garden hose and calibrated nozzle are needed for the spray test. To isolate points of water infiltration, a technician wets specific wall areas to see where leaks occur. The operator holds the nozzle about a foot from the wall and sprays water in a 5- to 10-foot-wide area, usually starting at the bottom of the wall. The time spent spraying the wall depends on the rainfall intensity for that region. The water pressure and flow rate through the nozzle can be calibrated to achieve any prescribed rainfall intensity and wind pressure. During the test, the interior of the wall is checked for water leakage. Significant leaks usually can be detected within 20 to 30 minutes.

When plastic sheeting is used to isolate certain wall locations, the test is called mask and spray. The test, which can be completed in about

Figure 10-2. To test the water permeance of mortar joints, the operator clamps this device to the wall and forces a known quantity of water into the wall. The water penetrates poor quality joints much faster than it does good quality joints.

an hour, is especially useful for checking water penetration through caulking, flashing, and expansion joints. It's a modified version of the Architectural Aluminum Manufacturers Association's "Field Check of Metal Curtain Walls for Water Leakage"[2].

10.4 THE PERMEABILITY TEST

A field technician uses a special device to test the water permeance of mortar joints or masonry walls (Figure 10-2). The operator drives two bolts into a bed joint and installs clamps that support the device on the wall. Water is pumped into the device until it reaches the 0 level in a graduated vertical tube. The operator selects the desired pressure and the water is forced into the wall at that pressure. The time required to force the known volume of water into the wall is recorded. The water penetrates poor quality joints much faster than it does good quality joints. A complete test can be done in 15 minutes and 30 tests can be done in a day.

Figure 10-3. A modified version of the ASTM E 514 laboratory test can be used to evaluate the water penetration through existing masonry walls. Water is sprayed into a chamber mounted on the wall and the amount of water leaking into the wall is measured. The wall is well designed and built if the leakage rate is less than a half gallon per hour.

10.5 MODIFIED ASTM E 514 FIELD TEST

ASTM E 514, *Test Method for Water Permeance of Masonry* [3], is a laboratory test modified for use on existing walls. A four-by-three-foot test chamber is mounted and sealed to the masonry wall (Figure 10-3). Water is pumped from a calibrated tank (supported at ground level) to a spray bar at the top of the chamber. After the wall is preconditioned for 30 minutes so that it is saturated with water, the water level in the tank is measured.

After the wall is preconditioned, water is sprayed on the wall for 4 to 8 hours. Tank water level is recorded every 30 minutes. The test is stopped when two consecutive water readings are the same. If this never occurs, the test is stopped after eight hours. The quantity of water lost from the tank in the last hour of the test is measured as the water leakagerate.

If the leakage rate is less than a half gallon per hour, the materials, bond, and workmanship are considered good [4]. Materials and workmanship may be questionable if the leakage rate is a half to 1 gallon per hour. A leakage rate greater than 1 gallon per hour indicates serious problems with materials or workmanship. The test can be performed in a day; set up takes 2 or 3 hours.

Figure 10-4. The drainage test can evaluate the drainage capacity of 40 feet of masonry cavity and flashing at one time. Through holes drilled in the exterior brick, water trickles into the cavity at rates based upon water penetration measured by a modified ASTM E 514 test procedure.

10.6 *THE DRAINAGE TEST*

A recently proposed test [5] evaluates the drainage of masonry cavities and flashings by introducing water directly into the cavity. Unlike ASTM E 514, which tests only a small area of flashing, this drainage test has been used to test as much as 40 feet of cavity drainage system at once.

A series of 3/8-inch-diameter holes are drilled through the masonry at spacings less than 24 inches on center. Water is pumped from a tank by a small hose with an in-line flow regulator and a flow meter. The regulator controls the water flow through each hole so the water drips, instead of squirts, into the cavity (Figure 10-4). The water flow rate should match the penetration rate measured by the modified ASTM E 514 test. During the test, the wall is continuously checked for water leakage. If necessary, openings are made into the wall to see how much water is entering the backup.

The total volume of water entering the wall and exiting the flashing and weep holes also can be measured. A temporary gutter is taped directly below the flashing and weep holes to collect the water and drain it into the tank. After about one hour of testing, the amount of water exiting the wall should be about the same as the amount entering the wall. If not, there may be interior leakage, requiring openings to determine the cause.

Test set-up time is 1 to 2 hours, followed by 1 to 2 hours of testing time. Four or five lengths of flashing as long as 40 feet can be tested in one day.

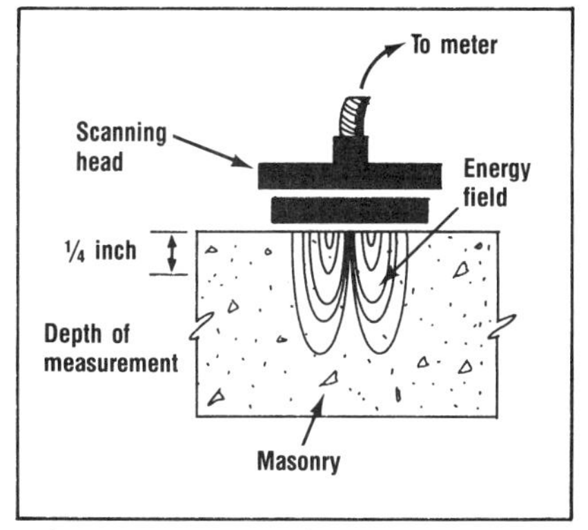

Figure 10-5. The meter works through a battery connected to two electrode plates generating an energy field in the masonry wall. The meter senses moisture in masonry by sensing the change in the energy field caused by the moisture. The meter only reads the moisture content in the top quarter inch.

10.7 *TESTING SAMPLE PANELS*

Before constructing a building, contractors—for their own protection—should consider having the water permeance of a test panel measured using the modified ASTM E 514 test. This will document what water permeance the contractor's workmanship provides. If the water permeance of the test panel is greater than what the architect specified, the test panel should be rebuilt to meet the architect's specification.

A test of the completed structure will indicate what water permeance the workmanship and building design provide. If the walls leak, but the test panel didn't, the building design may be at fault.

10.8 *MEASURE THE MOISTURE INSTANTLY*

Too much water in a masonry wall can cause all sorts of problems, including cracking, crazing, spalling, disintegration, efflorescence, decreased insulating efficiency, and deterioration of paints, plasters, or other finish-

Figure 10-6. Used as a diagnostic tool, the meter can help locate the cause of moisture in walls. The portable meter weighs about two pounds.

es. When stains appear on wall surfaces, water penetration is easy to detect, but it has already caused damage. Catching water penetration early, so it can be stopped before it causes damage, has been difficult.

However, a portable moisture meter makes early detection easier because a material's moisture content can be measured instantly with the electronic device. By taking readings at different spots on a wall, a profile of the wall's moisture content can be created in minutes. A high reading in one spot locates the main source of moisture penetration, which is also where repairs would be the most cost-effective.

Introduced in the United States in 1988, the portable moisture meter weights about 2 pounds. Unlike electrical resistance moisture meters, which measure the moisture content between two probes put into or on a wall, the meter measures the moisture content over an area. This reduces variations in readings caused by variations in the material. The flat scanning head of the meter is pressed against the wall, not into it. Some meters also emit an audio signal, allowing moisture content to be determined in complete darkness.

The meter measures moisture content by measuring the change in capacitance caused by water in the material. Capacitance exists between

Figure 10-7. A plastic sheet taped over a wall can help determine if moisture is moving in or out of the wall. If moisture readings over the plastic are higher than readings directly on the wall, then moisture is exiting the wall.

any two conductors separated by a dielectric (insulator). The dielectric may be air, masonry, concrete, or any other material that acts as an insulator. The circuit this dielectric completes is 81 times more sensitive to moisture than to air. As the moisture content of the material changes, it capacitance changes. The meter senses this change in capacitance and converts it into a moisture content reading.

Unfortunately, this electronic moisture meter gives more information than a masonry investigator can use. Though the meter can instantly find the exact moisture content of a masonry wall, it can't instantly be determined if that moisture content is harmful. The critical moisture contents for all masonry materials aren't known.

10.9 WHERE IS THE MOISTURE COMING FROM?

Despite this lack of data, the new meter can help in investigating moisture problems. To determine where moisture in a wall is coming from, a comparison of moisture contents of different areas of the wall must be done.

10.9.1 Rain

Moisture contents from wind-driven rain should be higher on the outside of walls than on the inside. If moisture readings increase near mortar joints, it's a sign that the moisture in the wall is from rain. Most rainwater penetration occurs through mortar joints, not through the masonry units.

To substantiate meter readings that indicate rainwater penetration, the workmanship of the wall and the driving rain index for the area must be examined. The driving rain index (with values from 1 to 5) is an indicator of wind pressure and annual precipitation [6].

10.9.2 Groundwater

Walls that are drawing groundwater by capillary action must have a moisture content at or above a critical water content, which for brick is about 6% [7]. Capillary action, or suction, can't occur if the water content is below this level. Meter readings also should be higher near ground level.

To confirm that the wall is drawing groundwater, check to make sure that the ground is damp from a high water table or continually heavy rainfall. A horizontal efflorescence line may also be visible near the base of the wall where salts are deposited during evaporation.

10.9.3 Condensation

Water probably is condensing inside a cavity wall if meter readings inside the cavity are higher than readings on the interior or exterior sides of the wall, and if the readings are taken in the upper half of the wall (to rule out groundwater). If temperature and relative humidity are such that the dew point is not reached, condensation will not occur.

10.10 IS THE MOISTURE ENTERING OR EXITING?

The meter measures only moisture content and not whether moisture is moving in or out of the wall. To find this out, a 2-by-2-foot plastic sheet should be taped over the wall. After four hours have passed, preferably during the night, six moisture readings can be taken at two locations: three readings over the plastic and three readings a couple feet away on the wall's surface. The three readings should be averaged. If the average reading over the plastic is higher than the average reading on the wall, moisture is exiting the wall. If the average reading over the plastic is lower than that of the wall, moisture is entering the wall.

Water trapped behind paints or other nonbreathable coatings can damage the paints or coatings. The plastic sheet acts like a nonbreathable coating and can be used to determine when painting or new construction can begin.

The moisture meter also can be used to determine if a wall that has been rained on has dried. If the same moisture reading is obtained a few days apart, the wall has reached a moisture content in equilibrium with its environment—and thus it's ready for painting or coating.

10.11 *A CHECK FOR GROUTED CORES*

The moisture meter is still being tested and new applications are being explored. At the University of Colorado, preliminary tests have been conducted to determine its potential for checking the quality of high-lift grouting. The meter is placed against the outside of the cells to measure the water that the units have absorbed from the grout. The moisture meter can read this increase in moisture content and confirm that the cells have been grouted. Although initial results are promising, more work is underway.

10.12 REFERENCES

1. RILEM. *Measurement of Water Absorption Under Low Pressure.* Paris: International Union of Testing and Research Laboratories for Materials and Structures.
2. American Architectural Manufacturers Association. 1983. *Field Checks of Metal Curtain Walls for Water Leakage,* 501.2-83. Des Plaines, Ill.: AAMA.
3. ASTM. 1990. *Test for Water Permeance of Masonry,* ASTM E 514 90. Philadelphia: ASTM.
4. Rath, Charles H. 1985. Brick Masonry Wall Nonperformance. *Masonry: Research, Application, and Problems,* ASTM STP 871. J.C. Grogan and J.T. Conway, eds. Philadelphia: ASTM.
5. Krogstad, Norbert V. 1990. Masonry Wall Drainage Test—A Proposed Method for Field Evaluation of Masonry Cavity Walls for Resistance to Water Leakage. *Masonry Components to Assemblages,* ASTM STP 1063. John H. Matthys, ed. Phildelphia: ASTM.
6. Grimm, Clayford T. 1982. A Driving Rain Index for Masonry Walls. *Masonry: Materials, Properties, and Performance,* STP 778. Philadelphia: ASTM.
7. Smith, Baird M. 1990. Moisture Problems in Historic Masonry Walls. National Park Service, Preservation Assistance Division, Superintendent of Documents. Washington, D.C.: U.S. Government Printing Office.

Chapter Eleven • Miscellaneous Techniques

Several additional techniques for evaluation of existing masonry buildings are described in this chapter. Many of these methods are in the process of being developed and show some promise for masonry evaluation. These techniques must be approached with caution: just because a method is based on the latest technology and uses high-tech equipment does not mean it is automatically applicable to masonry evaluation. It is important that people considering use of these methods understand the applications and limitations of the tests before using them for building evaluations.

11.1 *BOREHOLE DILATOMETER*

Determining engineering properties of materials on the interior of massive masonry structures is difficult and often requires partial dismantlement for access to inspect and test interior wythes. A small dilatometer, inserted into a borehole drilled through the structure, can be used to determine the deformability of interior wythes (Figure 11-1, adopted from Rossi [1]). The dilatometer uses an expanding sleeve to load the masonry around the borehole. Corresponding deformations are measured using internal instrumentation, and measured load-deformation response is used to indicate masonry deformability properties. Tests conducted at different positions within the borehole provide information on the variation of deformability between the exterior and interior layers. This technique was used successfully in Italy to prove the effectiveness of repairs to the foundation ring of the Tower of Pisa [1].

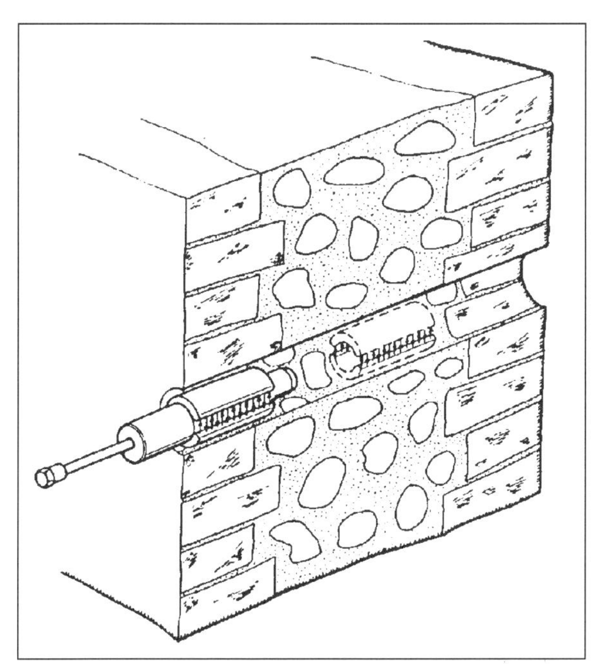

Figure 11-1. A borehole dilatometer is used to determine deformability properties of both the surface and inaccessible inner layers.

11.2 *LONG-TERM MONITORING*

It is sometimes desirable to monitor structural behavior under normal service loads. Masonry structures showing signs of distress can be instrumented to determine if deterioration is ongoing or static. Such information is critical for evaluating the safety of the structure under service loads and also is useful for monitoring the effects of restoration or retrofit work.

One of the more beneficial applications of long-term monitoring is determining the causes of observed distress. Crack openings may show daily fluctuations related to thermal movements; crack openings also may be affected by moisture, wind speed and wind direction. Environmental monitoring is almost always conducted as part of a long-term instrumentation program.

Instrumentation is used to monitor crack openings, horizontal movement, rotation of columns and walls, applied loads, environmental considerations such as humidity and temperature, and behavior of foundations including soil and rock. A wide variety of crack monitors, tiltmeters, load and pressure cells, strain measurement devices, and temperature probes are available for such work. The instruments usually are connected to automatic data acquisition and recording devices and also can be attached via phone line to a central recording station. The central computer can respond quickly to any irregularities in measured structural response and send notice if safety criteria are exceeded. Many historical and critical structures have been instrumented in such a manner.

11.3 *IMPACT ECHO*

The impact echo technique uses reflections of transient stress waves within a material to locate internal boundaries. Unlike techniques based on through-wall passage of waves and resultant velocity determinations, the

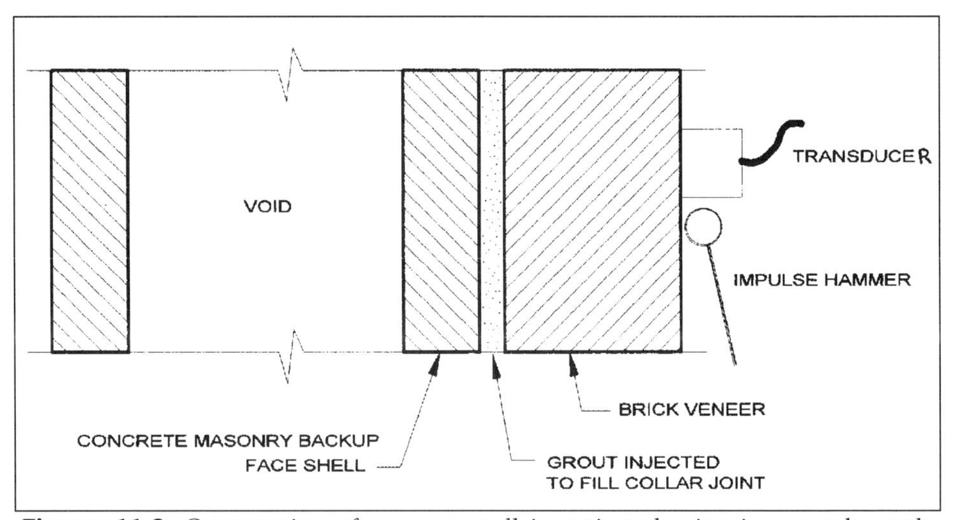

Figure 11-2. Cross-section of masonry wall investigated using impact-echo technique. Tests were used to determine if the collar joint was filled successfully with grout during a remediation procedure involving grout injection.

impact echo technique requires access to only one surface of the section being investigated. As seen in Figure 11-2, the impact source is located on the same surface as the receiving transducer. Cracks, voids, and interfaces between dissimilar materials can be located by analyzing the reflections of a stress wave off these boundaries. The method is used on concrete structures [3] and it is beginning to be used for masonry evaluation.

A transient stress wave is introduced into the material by any of a number of sources. A short-duration mechanical impact provides a crisp input signal and, by choosing impact sources with different contact times, the frequency content of the input signal can be optimized for a particular situation. The stress wave generated by such an impact is reflected off internal boundaries and the free edges of the specimen. Analysis of the received waveform in the frequency domain provides information on the time between successive reflections of the stress wave. When the pulse velocity of the material being investigated is known, the depth to internal flaws or the back surface of the section can be calculated using reflection time information.

11.3.1 *Impact Echo for Diagnosis and Control of Masonry Repairs*

Impact echo tests were used during a remediation project on a masonry building to locate internal voids and to verify the proper filling of voids by grout injection. Exterior walls of the structure consisted of a load-bearing backup wythe of hollow concrete masonry faced with a veneer of clay brick. The two wythes were separated by a narrow cavity with average thickness of 3/4 inch (Figure 11-2). Injection of grout into the partially empty collar joint was used to correct water penetration problems and structural deficiencies resulting from inadequate construction.

Impact echo tests were chosen for testing because access to only one surface was needed, the technique is non-destructive, and through-wall pulse velocity testing would have been complicated by the empty vertical cell in the interior CMU wythe. Results from impact echo testing at ungrouted and grouted areas are shown in Figures 11-3 and 11-4. When the collar joint is empty, the free surface at the back of the brick reflects the stress wave; 12,375 reflections per second are recorded as indicated by the dominant peak in the frequency spectrum at this level in Figure 11-3. This peak is reduced in the plot of Figure 11-4, indicating good bond between grout and brick, and is replaced by peaks at 11125 and 7750 Hertz. These frequencies correspond to reflections off the interface between the grout and the front surface of the concrete masonry unit (11125 Hertz) and the back of the CMU face shell (7750 Hertz). Spot testing was conducted at 8 inch centers in selected areas based on a random sampling procedure to verify proper filling of the collar joint by injected grout. Areas which were not fully injected or had poor bonding of the grout to brick surfaces were first inspected by borescope and remedial injection conducted.

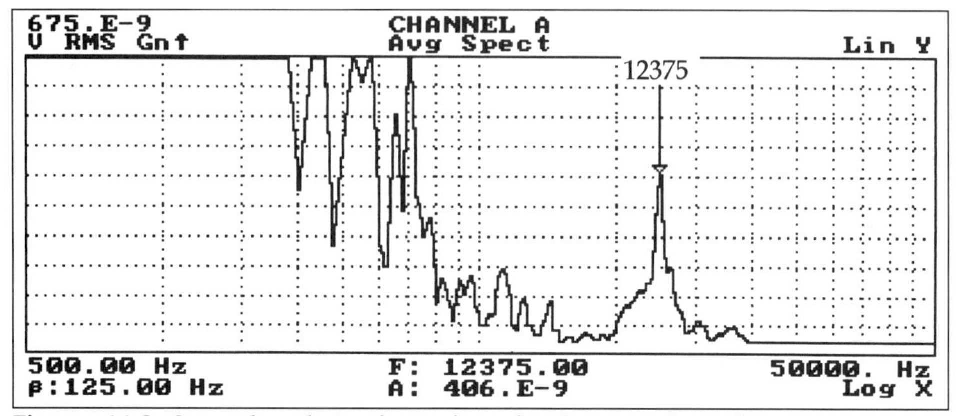

Figure 11-3. Spectral analysis of waveform for the case where the collar joint was ungrouted. The dominant frequency at 12375 Hertz corresponds to stress wave reflections off the back of the brick at the collar joint void.

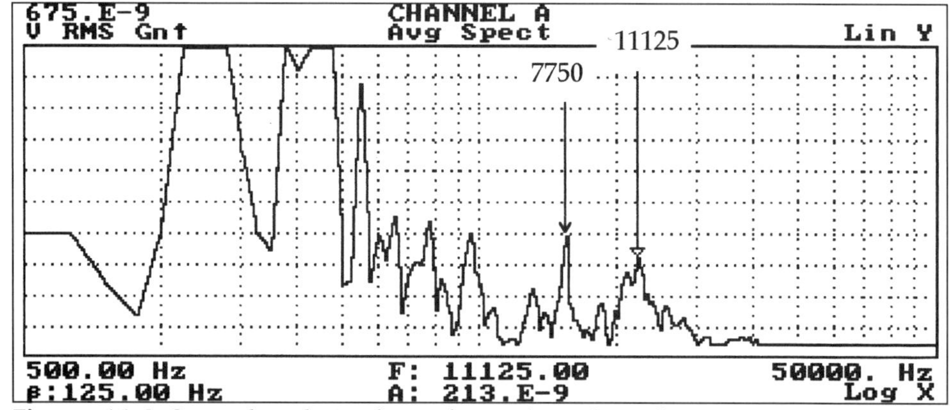

Figure 11-4. Spectral analysis of waveform where the collar joint was filled solidly with grout. Reflections off the back of the brick (12375 Hertz) have been reduced and instead, reflections at the CMU/grout interface (11125 Hertz) and the back of the CMU face shell (7750 Hertz) dominate.

11.4 *TOMOGRAPHIC IMAGING*

Tomography is a method for mathematically combining large amounts of projection data taken along many different ray paths as shown in Figure 11-5. The projections are used to reconstruct a cross-sectional image of the object being studied. The principle of reconstructing a function from its projections was first published by Radon in 1917 [2]. The method was not fully applied until the invention of the X-ray computed tomographic scanner; Hounsfield and Cormack shared a Nobel prize for this effort in 1972.

The use of computed tomography for imaging purposes has grown considerably, with many types of commercial equipment developed for

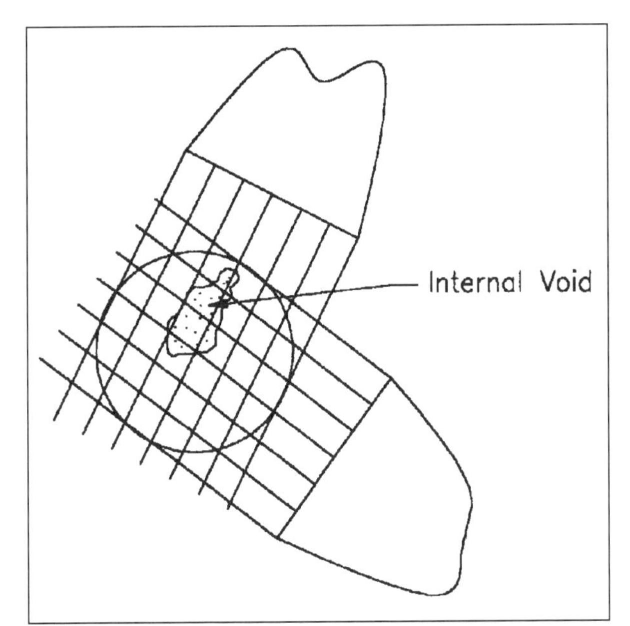

Figure 11-5. Projection of an internal anomaly from two directions. Tomographic reconstructions use a large number of such projections to develop an internal image of the object.

biomedical, electronic, and aerospace applications. The equipment uses a variety of techniques: X-ray radiography, nuclear magnetic resonance (NMR), biomagnetic computed tomography (BCT), radar, and acoustical imaging are all used for tomographic applications.

A crude form of tomography using acoustic pulse velocity is readily adaptable to field evaluation of masonry buildings. Acoustic pulse velocity data for tomographic image reconstruction are acquired by transmitting a high-frequency stress wave through the masonry through a number of different ray paths. The velocity of the stress wave varies as it passes through areas with different densities. In cases of voids or cracks, the pulse does not pass through the anomaly but seeks an alternate path around the flaw,

Figure 11-6. Masonry pier investigated using ultrasonic tomography as part of a project investigating repair of old masonry by injection grouting.

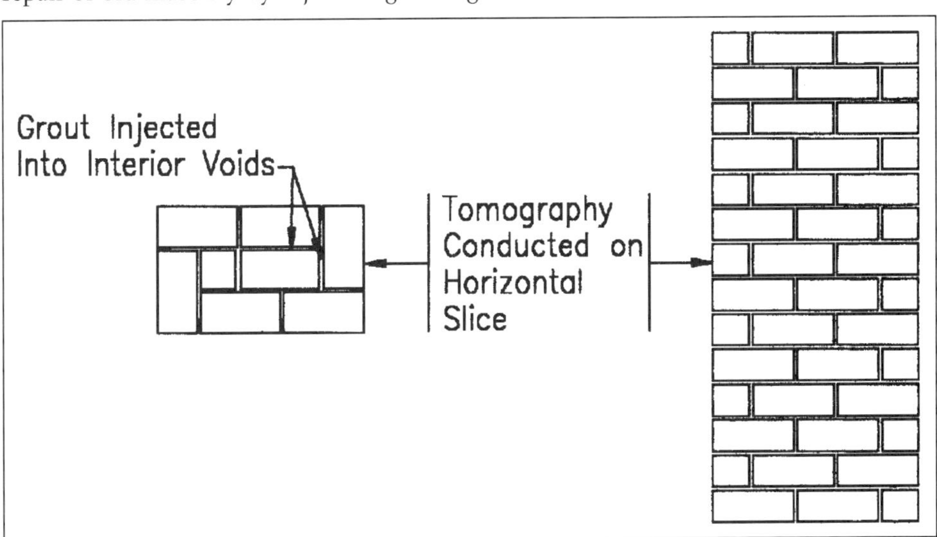

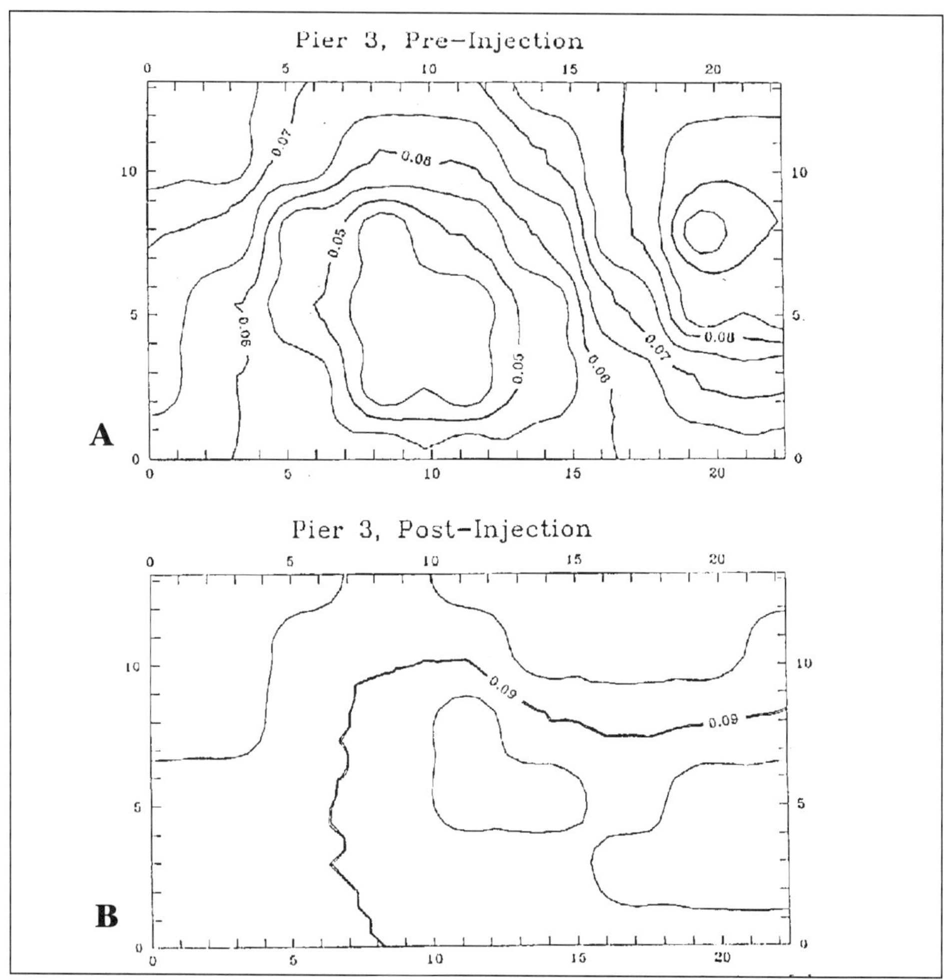

Figure 11-7 a & b. Tomographic reconstruction of internal velocity distribution of the masonry pier shown in Figure 11-3. (a) Original, as-built condition with interior voids. (b) Strengthened by injection grouting. Velocities shown are inches/μ sec.

resulting in an apparent reduction in velocity. Computer software is used to reconstruct an image of the internal velocity distribution based on the acquired data. The technique is appealing because it provides an accurate non-destructive technique for evaluating the internal condition of masonry using only surface measurements.

The reconstruction process is complicated by the scattering effect that damaged masonry has on incident stress waves. Furthermore, refraction and attenuation of high-frequency stress waves requires advanced analytical techniques for image reconstruction and even with such techniques the method produces only gross representations of the internal velocity distribution. The method also requires a large number of velocity readings to obtain an accurate reconstruction, a process that can be expensive and time consuming in the field.

11.4.1 *Imaging during Grout Injection*

A preliminary study of tomographic imaging for monitoring of grout injection into masonry was conducted by Italian and American researchers [4]. The method was successful for locating voids on the interior of a masonry pier before repair (Figure 11-6). Following repair by injection of grout into the masonry, tomography was used to verify the voids had been filled with grout. The contour plots in Figure 11-7 (a) and (b) show the variation in pulse velocity for a horizontal slice through the masonry pier for both the original and injected cases. The original pier was constructed with many internal voids approximately 3/8 inch across; these voids are evident by the low velocities in the interior of the specimen in Figure 11-7 (a). The post-injection plot of Figure 11-7 (b) shows an overall increase in pulse velocity, indicating that the voids were successfully injected with grout. Hence the technique is proven to be successful in locating overall velocity differences but was not able to resolve small individual voids.

11.5 *BORESCOPE INVESTIGATIONS*

Many situations require a visual investigation for determining the condition of hidden objects, or for inspecting cavities. A borescope is a simple device inserted into a hole for internal inspection of masonry walls (Figure 11-8). The borescope unit incorporates fiber optics and an internal light source to illuminate the cavity being inspected. Some borescopes have a graduated scale on the viewfinder to aid in identification and sizing of objects.

Figure 11-8. A borescope is a useful tool for investigating cavities. A small diameter hole is drilled in a mortar joint; an internal light source illuminates the cavity.

Borescopic investigations are cost effective and do not require localized dismantlement to inspect wall interiors. A 1/2 inch diameter hole is usually sufficient for insertion of borescope probes. When possible, holes should be drilled at mid-height of head joints to avoid mortar blockages which can be present at bed joint locations. Potential applications include:

- Inspection of collar joint blockage and mortar coverage in multiwythe construction
- Determination of materials and construction quality of interior wythes

- Location and inspection of veneer ties
- Corrosion inspection for masonry veneer ties, shelf angles, and steel stud systems
- Cavity inspection for cavity wall construction
- Inspection of weep hole blockages and flashing systems
- Inspection of internal cracks and voids located using other nondestructive techniques

Optional camera and video attachments attach to most borescopes and provide a means to obtain a permanent record of borescopic investigations.

11.6 ACOUSTIC EMISSION

Measurement of acoustic emissions from stressed materials is used to locate cracks as they form, determine flaw location and severity, and assess the integrity of masonry structures. The technique can be nondestructive, relying on service loads or ultrasonic pulses to generate emissions, or may use externally applied loads from hydraulic cylinders or dead weights. Acoustic emission monitoring has been applied to laboratory investigations for determining the integrity of masonry [5].

Strains develop in masonry when it is subjected to compressive, tensile, or shear stresses. This strain energy is released with the formation of microcracks at stresses which can be as little as 10% of the peak strength. As cracks propagate, a small amount of energy is released as transient stress waves. Sensitive accelerometers are used to detect this energy and, if multiple sensors are used along with a timing circuit, it is possible to locate the point of origination or source of the energy in three-dimensional space. Acoustic emission measurement is powerful in this respect but also is used for determining prior peak loading levels, stress levels initiating damage, the severity of the damage being caused, low-cycle fatigue cracking, and identification of long-term creep under sustained loads.

11.7 INFRARED THERMOGRAPHY

Infrared imaging provides a high-resolution heat picture of the surface of a masonry structure, allowing for rapid assessment of general condition. The method does not require direct access to walls under investigation and large structures can be investigated remotely from the ground. Sound areas of masonry affect heat flow differently than damaged or deteriorated areas, which may contain voids, cracks, or moisture. Surface temperatures are measured by a portable infrared scanning camera, which is often attached to computerized equipment for image enhancement. The two-dimensional color or gray-scale images can be recorded on videotape for future comparisons to determine the progress of deterioration or the effectiveness of repairs.

A heat source is required to drive heat flow from one side of a wall to the other. Heating may be passive, using heat radiation from the sun or earth, or actively driven using internal building temperature. Thermographic measurements are affected by many conditions, including wind, moisture, cloud cover, and surface condition, in addition to internal masonry condition.

This technique has proven to be quite capable for locating delamination cracks in reinforced concrete [6], and subsurface deterioration of pavements [7,8,9] with a minimum identifiable flaw size of 4 inches in diameter. A greater thermal gradient increases the resolution: deVekey reports it is possible to identify masonry ties in a filled cavity with a temperature gradient of 50 degrees C [10]. Relatively large internal voids can be located quickly using thermography, and the technique would be useful for locating voids in solid masonry, locating mortar droppings blocking weep holes [9], and delineating grouted and ungrouted areas in reinforced masonry construction. The method does not provide any real quantitative information on masonry material properties but would be useful for rapidly evaluating large areas, and could quickly locate areas requiring further investigation.

11.8 *IMPULSE RADAR*

Pulses of microwave energy can be used to evaluate construction materials such as soil, concrete, and masonry. The technique uses a small antenna that emits a brief pulse of electromagnetic energy in the frequency range of 100 MHz to 1 GHz. A receiver, located on the opposite side of the mass being investigated, records the signal. This technique requires access to two opposing surfaces and would not be directly applicable in many situations. Recent advances in reflected radar techniques appear to be more promising for general evaluation purposes. In this technique, the wave is reflected at boundaries between two materials with different dielectric properties and recorded by a receiver located next to the transmitter. The reflected wave is recorded and analyzed to determine the depth of the interface. This entire process is repeated at a frequency of 50 kHz, allowing a continuous record of subsurface conditions to be obtained as the antenna passes over the masonry surface. Wave transmission is affected by materials with a high moisture content and also by dissolved salts, complicating interpretation of the data.

Impulse or ground penetrating radar (GPR) was originally developed for geophysical applications but also has been used for the evaluation of pavements [6,8] and masonry [11,12] and was found to be 90% accurate in locating 1/4-inch-deep voids and honeycombs in concrete [6]. Impulse radar can be used to locate subsurface voids, piping, reinforcing bars, and damp areas. Recently impulse radar has been applied to the evaluation of masonry in the United States [12] and was found to be 95% effective in locating ungrouted cavities in concrete masonry walls. Application of impulse radar to global masonry evaluation would provide a method offering rapid assessment of masonry condition for locating areas needing further investigation.

11.9 *NEUTRON-GAMMA RADIOGRAPHY*

A radiographic technique using high-intensity neutron irradiation provides a method for determining the elemental composition of masonry building materials [13]. Neutrons interact with masonry materials to produce gamma rays and the gamma ray intensity, depending on the elemental concentration of the material. The technique provides a localized reading over a hemisphere approximately 6 to 8 inches in diameter adjacent to the neutron source. A number of discrete readings are required for characterization of large surface areas.

Neutron-gamma radiography is used to locate internal steel and voids and to determine salt and moisture profiles over the surface. The gamma ray intensity spectra have discrete peaks for elements such as silicon, hydrogen, calcium, chloride, and iron. Elemental concentrations are calculated based on the intensity of the spectral peaks, providing a method for determining water content and chloride profiles and locating reinforcing bars. The equipment required for such an investigation is readily available but highly specialized, with a cost of around $100,00 for portable equipment adaptable to field usage. Skilled electrical technicians and equipment operators are required to assemble and maintain the equipment. An average survey of building materials would cost several thousand dollars. The neutron source, although it emits minimal radioactivity, is a health and safety hazard to personnel. Before conducting a neutron-gamma test, licenses are required for the equipment and for the technician from the Nuclear Regulatory Commission. In spite of the cost and complexity of neutron-gamma radiography, the technique is totally nondestructive and provides information not easily obtained using alternate methods.

11.10 *X-RADIOGRAPHY*

X-radiography provides another nondestructive method for examining the internal condition of masonry. The method is used for inspecting construction quality and locating voids or deterioration on the masonry interior. A high-energy X-ray source is used to illuminate the test specimen from one side; a photographic film plate is located directly opposite the source to record the image. Beam transmission is affected by materials in its path: masonry units, mortar, steel, and air voids all scatter and attenuate the beam energy in a different way. X-ray equipment is expensive and difficult to operate in the field. High voltage and high radiation levels require operation by trained personnel. Recent advances in portability of X-ray generators may lead to more extensive use of this technique in the future.

Cracks, voids, reinforcing bars, and inadequately consolidated grout all can be identified with X-ray views. The technique may not identify small cracks or defects perpendicular to the X-ray beam. In addition, each view is representative of a condition throughout the entire masonry thickness at the test area and requires careful interpretation for locating the depth of features exhibited in the view. Radiographic views are obtained rapidly and require from less than one second to 30 minutes exposure time, depending on the source energy and mass being investigated. X-radiography has been used on concrete with a thickness of up to 36 inches; its use for masonry evaluation has been limited to date.

11.11 *LASER INTERFEROMETRY*

Holographic interferometry and speckle interferometry have been used to monitor fracture and debonding during tests of laboratory specimens [14]. Fringe or speckle patterns are formed by interference of laser beams projected onto the test specimen. Cracks and deformations form abrupt breaks or changes in fringe and speckle patterns as the specimen is loaded. The technique is intriguing for monitoring building deflections because it offers a method for global deformation monitoring, rather than the extremely localized strain measurements now used. Field applications would include location of strain concentrations, determining residual stresses following cutting or drilling operations, monitoring deformations during load tests, determining vibrational mode shapes of large and complex structures, and providing real-time monitoring of crack initiation and growth.

11.12 *PETROGRAPHY*

How can the mortar color of an existing building be matched? Does the mortar contain the mix proportions specified? Why is the mortar crumbling?

Once mortar is in the wall, these questions can still be answered. Chemists and petrographers can perform tests on hardened mortar to find out the amount and types of ingredients in it. Although these tests aren't routinely performed, architects, engineers, and contractors should be aware of them.

11.12.1 *What's in the Mortar?*

Typical mortar ingredients include sand, hydrated lime, masonry cement, portland cement, fly ash, blast-furnace slag, and limestone. Using a microscope, petrographers can identify the mineralogical composition and gradation of the sand and the type of cementitious materials used. Petrographers use ASTM C 295, *Petrographic Examination of Aggregate for Concrete,* and ASTM C 856, *Petrographic Examination of Hardened Concrete.* Following ASTM C 85, *Cement Content of Hardened Portland Cement Concrete,* chemists can determine the amount of cementitious materials used.

The percentages of mortar constituents are determined by weight but can easily be converted to volume proportions and compared with those in ASTM C 270, *Standard Specification for Mortar for Unit Masonry.* Thus it can be determined whether the mortar meets the volume proportions required for a Type M, S, N, or O mortar.

11.12.2 *Cement*

Typically, Portland cement and lime or a masonry cement is the binding agent for mortar. Petrographic examination often can distinguish if hydrated lime has been added to the mortar. For a Portland cement and hydrated lime system, chemists use the amount of soluble silica in the mortar to determine the amount of Portland cement. The amount of hydrated lime is

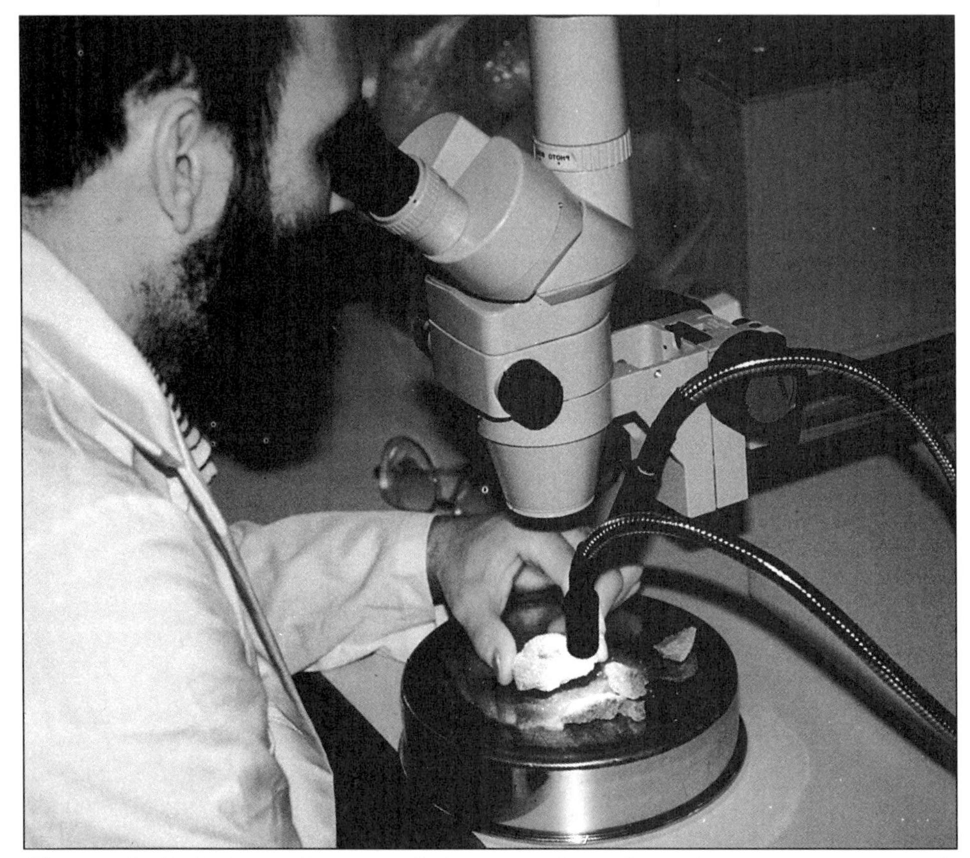

Figure 11-9. A stereomicroscope helps researchers determine aggregate composition, aggregate grading, air content, type of cementitious materials, and porosity in hardened mortar.

determined by the calcium contributed by the Portland cement. If, however, a limestone sand was used, the calculations will be skewed.

Determining the amount of masonry cement in the mortar may be more difficult. Different masonry cement manufacturers add slightly different ingredients and in varying amounts. Most masonry cements, however, are composed of Portland cement and ground limestone.

The amount of Portland cement is again determined by measuring the soluble silica in the mortar. Because masonry cements are about one-half Portland cement, the amount of masonry cement in the mortar is simply twice the Portland cement content.

11.12.3 *Sand*

Once the quantity of cementitious materials is known, the sand content can be calculated by subtracting the weight of the cementitious materials and moisture from the total weight of the mortar.

11.12.4 *Admixtures*

Air-entraining agents, integral waterproofers, accelerators, polymers, and other organic admixtures in the mortar can be identified by using infrared spectrophotometric analysis. Solvents are used to separate organic compounds from the mortar. An infrared spectrum is then obtained on the separated fraction and compared with reference spectra for identification.

Figuring out the amount of each admixture is more difficult. After identifying the admixture, trial mortar batches are prepared with different dosage rates. The amount of organic compounds in each trial batch is measured. Then these amounts are compared with the amount in the field sample to figure out the amount of admixture in the hardened mortar.

11.12.5 *Composition Affects Durability*

Why is the mortar crumbling? A petrographic examination can provide the answers. Petrographers can determine the air content, evaluate cement hydration, and even estimate in place water-cement ratio. They also can visually judge the interface between mortar and masonry unit to determine the potential for water penetration and bond.

11.13 REFERENCES

1. Rossi, P.P. 1990. Nondestructive Evaluation of the Mechanical Characteristics of Masonry Structures. Proceedings at the Conference on Nondestructive Evaluation of Civil Structures and Materials, University of Colorado. Boulder, Colo.
2. Kak, A.C., and M. Slaney. 1988. *Principles of Computerized Tomographic Engineering.* New York: The Institute of Electrical and Electronics Engineers Press.
3. Sansalone, M., Y. Lin, D. Pratt, and C.C. Cheng. 1991. Advancements and New Applications in Impact-Echo Testing, ACI SP 128-9. Evaluation and Rehabilitation of Concrete Structures and Innovations in Design Proceedings at American Concrete International Conference, SP-128. Hong Kong.
4. Schuller, M., M. Berra, A. Fatticcioni, R. Atkinson, L. Binda. 1994. Use of Tomography for Diagnosis and Control of Masonry Repairs. Proceedings at the 10th International Brick/Block Masonry Conference. Calgary, Alberta, Canada.
5. Leaird, J.D. 1984. A Report on the Pulsed Acoustic Emission Technique Applied to Masonry. *Journal of Acoustic Emission.* Vol. 3, no.4.
6. Holt, F., and J. Eales. 1987. Nondestructive Evaluation of Pavements. *Concrete International.* June.
7. Cantor, T.R. 1984. Review of Penetrating Radar as Applied to Nondestructive Evaluation of Concrete, In Situ/Nondestructive Testing of Concrete, ACI SP-82.
8. Manning, D.G., and T. Masliwec. 1990. Operational Experience Using Radar and Thermography for Bridge Condition Survey. Proceedings at the conference on Nondestructive Evaluation of Civil Structures and Materials. Boulder, Colo.

9. Stockbridge, J.G. 1979. Evaluating the Condition of Masonry Buildings in Service. Proceedings of V International Brick Masonry Conference. Washington, D.C.
10. de Vekey, R.C. 1988. Nondestructive Test Methods for Masonry Structures. Proceedings of the 8th International Brick/Block Masonry Conference. Dublin, Ireland.
11. Rossi, P.P. 1990. Nondestructive Evaluation of the Mechanical Characteristics of Masonry Structures. Proceedings at the Conference on Nondestructive Evaluation of Civil Structures and Materials. Boulder, Colo.
12. Lim, M.K., and C.A. Olson. 1990. Use of Nondestructive Impulse Radar in Evaluating Civil Engineering Structures. Proceedings at the Conference on Nondestructive Evaluation of Civil Structures and Materials. Boulder, Colo.
13. Livingston, R.A., L.G. Evans, T.H. Taylor, and J.I. Tomoka. 1986. Diagnosis of Building Condition by Neutron Gamma Ray Technique. *Building Performance: Function, Preservation, and Rehabilitation,* ASTM STP 901. Philadelphia: ASTM.
14. Maji, A.K. 1990. Civil Engineering Applications of Laser Interferometry. Proceedings at the Conference on Nondestructive Evaluation of Civil Structures and Materials. University of Colorado, Boulder, Colo.
15. Malhotra, V.M., and N.J.Carino, eds. 1990. *Handbook on Nondestructive Testing of Concrete.* Boca Raton, Flor.: CRC Press.
16. Bungey, J.H. 1982. *The Testing of Concrete in Structures.* New York: Chapman and Hall.
17. Clifton, J.R. 1985. Nondestructive Evaluation in Rehabilitation and Preservation of Concrete and Masonry Materials, ACI SP-85. *Rehabilitation, Renovation, and Preservation of Concrete and Masonry Structures.* Detroit: American Concrete Institute.
18. Modena, C., P.P. Rossi, and C. Bettio. 1991. Diagnosis and Strengthening of an Ancient Bell-Tower. Proceedings at the 9th International Brick/Block Masonry Conference. Berlin, Germany.
19. Atkinson, R.H., M.P. Schuller, and J.L. Noland. 1991. Nondestructive Test Methods Applied to Historic Masonry Structures. Proceedings at the Seismic Retrofit of Historic Buildings Conference. San Francisco.
20. Abrams, D.P. and J.H. Matthys. 1991. Present and Future Techniques for Nondestructive Evaluation of Masonry Structures. *The Masonry Society Journal.* Vol. 10, no.1.
21. Livingston, R.A. 1985. X-ray Analysis of Brick Cores from the Powell-Waller Smokehouse, Colonial Williamsburg. Proceedings at the Third North American Masonry Conference. Arlington, Texas.

• APPENDIX I •
Manufacturers and Suppliers of Masonry Test Equipment

There are many manufacturers and distributors of structural testing and evaluation equipment. The following listing is not intended to provide a complete inventory of test equipment manufacturers, but instead concentrates on providing sources for equipment applicable to laboratory and in-place masonry evaluation.

AGEMA INFRARED SYSTEMS — 201-867-5390
550 County Ave.
Secaucus, NJ 07094
Infrared thermography

AMBER ENGINEERING — 805-683-6621, Fax: 967-4208
5756 Thornwood Dr.
Goleta, CA 93117
Infrared thermography

AMERICAN SOCIETY FOR TESTING & MATERIALS — 215-299-5400, Fax: 299-2630
1916 Race Street
Philadelphia, PA 19103-1187

ATKINSON-NOLAND & ASSOCIATES INC. — 800-735-3629, 303-444-3620, Fax: 444-3239
2619 Spruce St.
Boulder, CO 80302
Flatjack testing equipment for masonry, custom fabrication of test fixtures and test equipment packages

AVONGARD — 312-244-4179, Fax: 244-6685
2836 Osage
Waukegan, IL 60087
Crack monitoring equipment

BLOUNT INC. — 503-653-8881
4909 S.E. International Way
Portland, OR 97222
Diamond-segmented chain saw

CC TECHNOLOGIES — 614-761-1214, Fax: 761-1633
2704 Sawbury Blvd.
Columbus, OH 43235

DIAMOND PRODUCTS
333 Prospect St.
Elryia, OH 44035
Dry cutting equipment for mortar removal, tuckpoint grinders, masonry saws
800-321-5336
216-323-4616
Fax: 323-8689

ECKERT OPTICAL INSTRUMENTS INC.
412 Halsey Rd.
Annapolis, MD 21401
Borescopes, microscopes, video inspection services
410-269-1415
Fax: 974-8290

ELE/SOILTEST INC.
86 Albrecht Dr.
P.O. Box 8004
Lake Bluff, IL 60044
Reinforcement location, Schmidt hammer, crack monitor, ultrasonic pulse velocity, laboratory equipment for material testing
800-323-1242
708-295-9400
Fax: 295-9414

FaAA PRODUCTS CORP.
149 Commonwealth Dr.
Menlo Park, CA 94025
Eddy-current, reinforcement location
415-688-7181
Fax: 326-8072

FORNEY INC.
Rt. 18, R.D. 2
Wampum, PA 16157
Schmidt hammer, reinforcement location, moisture measurement, laboratory test equipment
412-535-4341

GEOPHYSICAL SURVEY SYSTEMS INC.
15 Flagstone Dr.
Hudson, NH 03051
Subsurface interface radar
800-524-3011
603-889-4841
Fax: 889-3984

GERMANN INSTRUMENTS INC.
8845 Forest View Rd.
Evanston, IL 60203
Pulse-echo, pullout, pulloff, corrosion, reinforcement location, water permeability, ultrasonic, moisture content, crack depth, rapid chloride determination, carbonation, coring equipment
708-329-9999
Fax: 329-8888

HUMBOLDT MFG. CO. 708-456-6300
7300 W. Agatite Ave.
Norride, IL 60656
Ultrasonic pulse velocity, Schmidt hammer, reinforcement location, probe penetration, laboratory test equipment

INFRAMETRICS 617-275-4510
12 Oak Park Dr. Fax: 272-5207
Bedford, MA 01730
Infrared thermography

ISS THERMOGRAPHICS 410-875-0234
4757 Buffalo Rd.
Mount Airy, MD 21771
Infrared thermography

JAMES INSTRUMENTS INC. 800-426-6500
3727 North Kedzie Ave. 312-463-6565
Chicago, IL 60618 Fax: 463-0009
Ultrasonic pulse velocity, Windsor penetrometer, pin penetrometer, reinforcement location, pull-off, moisture meter

MATEC INSTRUMENTS 508-435-9039
75 South St. Fax: 435-2165
Hopkinton, MA 01748
Ultrasonic test equipment

NEWPORT CORP. 714-963-9811
18235 Mt. Baldy Circle
Fountain Valley, CA 92708
Holographic cameras for interferometry

OLYMPUS CORP. 516-488-3880
4 Nevada Dr.
Lake Success, NY 11042
Fiber-optic borescopes

PCB PIEZOTRONICS 716-684-0001
3425 Walden Ave. Fax: 684-0987
Depew, NY 14043
Accelorometers, power supplies, signal conditioners, and test hammers for sonic testing

PHYSICAL ACOUSTICS 609-844-0800
P.O. Box 3135 Fax: 895-9726
Princeton, NJ 08543
Acoustic emission test equipment

PRG 301-309-2222
Preservation Products Catalog FAX: 279-7885
P.O. Box 1768
Rockville, MD 20849
Moisture meters, hand tools, books, general equipment for evaluation, repair, and rehabilitation

RGC 716-895-1156
P.O. Box 681 Fax: 895-1547
Buffalo, NY 14240
Diamond-segmented chain saw, circular saw

RILEM
Secretariat General 33 1 47 40 23 97
12 Rue Brancion Fax: 33 1 47 40 01 13
75737 Paris CEDEX 15
FRANCE
European agency that produces test standards for load-bearing unit masonry

ROCTEST 518-561-3300
7 Pond St. Fax: 561-1192
Pittsburgh, NY 12901
Borehole dilatometer, tilt and strain meters

SCHLEUTER INSTRUMENTS CORP. 303-530-2217
4699 Nautilus Ct. Fax: 530-7694
Suite 105
Boulder, CO 80301
Fiber optic borescopes

SCHONBERG RADIATION CORP. 408-980-9729
3300 Keller St. Fax: 980-8605
Building 101
Santa Clara, CA 95054
High-energy X-ray equipment

SDS CO.
P.O. Box 844
Paso Robles, CA 93447
Structural movement monitoring,
moisture meter, Schmidt hammer,
rebar locator, ultrasonic test equipment,
crack width meter, pull-off

805-238-3229
Fax: 238-3496

WESTERN NDE & ENGINEERING LTD.
4220 Rossiter Dr.
Victoria, BC
V8N 4S7
CANADA
Ultrasonic pulse velocity testing equipment

604-477-7824
Fax: 477-7824

• Index •

V

W

X